Wild, Wild World of Animals

A TIME-LIFE TELEVISION BOOK

Produced in Association with Vineyard Books, Inc.

Editor: Eleanor Graves
Senior Consultant: Lucille Ogle
Text Editor: Richard Oulahan
Associate Text Editors: Bonnie Johnson, Jack Murphy, Milton Orshefsky
Advisory Editor: Bertel Bruun
Authors: Prue and John Napier
Assistant Editor: Regina Grant Hersey
Research: Ellen Schachter, Elsie Washington
Copy Editor: Robert J. Myer
Picture Editor: Richard O. Pollard
Picture Research: Judith Greene, Barbara Crosby
Permissions: Jo-Anne Cienski
Book Designer and Art Director: Jos. Trautwein
Creative Consultant: Philip H. Wootton Jr.
Production Coordinator: Jane Quinson

WILD, WILD WORLD OF ANIMALS
TELEVISION PROGRAM
Producers: Jonathan Donald and Lothar Wolff
This Time-Life Television Book is published by Time-Life Films, Inc.
Bruce L. Paisner, *President*
J. Nicoll Durrie, *Business Manager*

THE AUTHORS

PRUE NAPIER is the British Museum's expert on monkeys and apes. Before taking up her present post she worked at the Smithsonian Institution, Washington, D. C., where she assisted her husband, Professor John Napier, in setting up a Program of Primate Biology. She is co-author, with her husband, of a standard work on the primates, *A Handbook of Living Primates*, and several other books on the subject of monkeys and apes.

PROFESSOR JOHN NAPIER is an anatomist and anthropologist and a leading authority in Britain on the primates. He has a medical background and served as an orthopedic surgeon during World War II. At the end of the war he gave up medical practice in favor of research into primates and human evolution. He introduced the first teaching programs in primate biology in Britain. He is at present Visiting Professor at Birkbeck College, University of London.

THE CONSULTANTS

WILLIAM G. CONWAY, General Director of the New York Zoological Society, is an internationally known zoologist with a special interest in wildlife conservation. He is on the boards of a number of scientific and conservation organizations, including the U. S. Appeal of the World Wildlife Fund and the Cornell Laboratory of Ornithology. He is a past president of the American Association of Zoological Parks and Aquariums.

DR. JAMES W. WADDICK, Curator of Education of the New York Zoological Society, is a herpetologist, specializing in amphibians. He has written for many scientific journals and has participated in expeditions to Mexico, Central America and Ecuador. He is a member of the American Society of Ichthyologists and Herpetologists, a Fellow of the American Association of Zoological Parks and Aquariums and a member of its Public Education Committee.

JAMES G. DOHERTY, as curator of Mammals for the New York Zoological Society, supervises the mammal collection of approximately 1,000 specimens at the Society's Zoological Park in the Bronx, New York. He is the author of many articles on the natural history, captive breeding and management of mammals. He is a member of the American Association of Mammalogists and a Fellow of the American Association of Zoological Parks and Aquariums.

Wild, Wild World of Animals

Based on the television series
Wild, Wild World of Animals

Published by
TIME-LIFE FILMS

ISBN 0-913948-03-9

Library of Congress Catalog Card Number: 75-40056

Contents

Introduction

by Prue and John Napier

THE PRIMATES, a group that includes man, are so called because they are regarded as the leading mammals, with a more complex and advanced brain than any other animals. They are also leading mammals in quite another sense. They were among the earliest groups to separate from the original ancestral mammalian stock.

When the age of giant reptiles, the dinosaurs, came to an end some 65 million years ago, the mammals were very small, rather insignificant, insect-eating creatures living in a humid environment in which forests, as we know them today, did not exist. The principal plants of the time were ferns and conifers; a few flowering plants, heralding the new era of the great forests of the future, were scattered here and there, adding a touch of bright color to the universal greens and browns of the vegetation. Everywhere was water—lakes, swamps and inland seas.

Among the first mammals to emerge were the primates, in a time when the world was undergoing a period of major geological change: the Laramide Revolution, which lasted for several millions of years and left its mark on the shape of the land surface, the climate and the vegetational cover. Inevitably the balance of life in the animal world was disturbed. The dinosaurs vanished for good, and the mammals, the birds and the insects became the dominant forms of animal life. Forests of deciduous, flowering trees attracted insects, and insects attracted birds and mammals. Until that time of change, the mammals had been terrestrial, making their nests in burrows, in decaying logs and among the roots of trees.

Some 60 million years ago in the Paleocene Epoch, all the earliest mammals looked very much alike. They had long bodies, short limbs, small eyes and exceptionally long and mobile noses. There were few physical differences between insectivores (shrews, hedgehogs, etc.), primates or carnivores. They all had much the same diet of insects, grubs and small lizards and lived the same sort of lives. Then came the forests, and these early mammals began to evolve into definite, easily distinguished groups. The forest floor remained the home of insectivores and other small mammals, while the forest canopy was successfully colonized by the primates. With this change of habitat the primates had to adapt to new kinds of food. Instead of an exclusively insectivorous diet, tree products such as leaves, fruit, buds and flowers also figured in their menu. As the primates increased in size during evolution, insects were gradually replaced by a bulk vegetarian diet. Today only the smallest of the primates, such as the South American marmosets, still depend mainly on insects for their food.

Life in the trees rapidly changed the outward appearance of primates, and some 10 million years later in the Eocene Age there could be little doubt that the animals climbing, swinging and leaping among the branches of the forest canopy were the monkeys and apes of the future.

There are two main groups of living primates: the prosimians and the anthropoids, called "lower" and "higher" primates respectively. Prosimians, which, literally translated, means "pre-monkeys" or "early monkeys," represent the transition stage between the insectivores and monkeys. Found in Africa, Madagascar and Southeast Asia, the prosimian group is divided into six families: tree shrews (believed to be similar to the primitive mammalian stock from which the primates evolved), lemurs, indrises, aye-ayes, lorises and tarsiers. In this book we are concerned only with the higher primates, excluding man, which are divided into five families. They differ profoundly in their zoogeographical distribution, in size, in social organization and behavior and, above all, in intelligence. Two of the five families are from South America and three from Africa and Asia. Collectively, South American monkeys are often referred to as "New World Monkeys," in contrast to the monkeys from the rest of the world, such as the baboons, langurs and macaques, which are known as "Old World Monkeys." There are some fundamental differences between New and Old World forms that concern their size, their teeth, their noses, their thumbs, their tails and their sitting pads, or ischial callosities.

The Old World Monkeys form one very large and diverse family in which there are two major divisions called subfamilies. The first includes the omnivorous guenons, macaques, mangabeys and baboons. The second subfamily consists of the langurs of Southeast Asia and the colobus monkeys of Africa, a widespread and successful group that lives very largely on a diet of leaves. Colloquially they are referred to as the leaf-eaters, and, to accommodate their special diet, their digestive systems are quite different from those of omnivores. For example, the langurs have a modified stomach not unlike the stomach of cows and other grass-eating mammals.

The New World cebids and marmosets live almost entirely on foliage, seldom leave the trees and are generally smaller than their Old World cousins.

	NEW WORLD MONKEYS	OLD WORLD MONKEYS
Noses	Nostrils well separated, facing sideways	Nostrils close together, facing downward and outward
Thumbs	Not fully opposable	Where present, fully opposable, or capable of being placed opposite, as a thumb to forefinger
Tails	Sometimes prehensile	Never prehensile
Sitting pads	Absent	Present in many

Of all the primates, woolly monkeys, like the one above, as well as spider and howler monkeys, have the most highly specialized tails (apes, by definition, are tailless). These remarkable appendages, called prehensile tails, are superb grasping devices capable of supporting the monkeys' entire weight, thereby leaving the hands free for feeding. The end of the tail (right) is covered with bare skin marked with ridged "fingerprints," making the tail as sensitive as a fingertip.

One of the distinguishing features between Old and New World Monkeys is the different shape of their noses. The nostrils of the Old World Monkeys, like the pig-tailed macaque above, are usually elongated, point downward and are set close together. The New World Monkeys, such as the white-nosed saki at right, usually have round nostrils separated by a broad nasal septum.

Finally, there are two ape families: the lesser apes, including the gibbon and siamang, and the great apes—the gorilla, the chimpanzee and the orangutan. Despite the many differences that separate the five families of higher primates, there are certain characteristics they all share that distinguish them from other mammals.

Sense of smell: It is generally true to say that creatures living in trees depend less for their survival on a sense of smell than do groundlings. A highly developed olfactory system is important for predatory animals whether they are active at night or by day. Higher primates are not hunters of animal food, nor, with one exception—the South American douroucouli—are they active in the dark.

The sense of smell in higher primates has become much reduced in importance since their early evolutionary days, and in consequence their noses have become less complex and their muzzles shorter. Thus the majority of primates have rather flat faces. One of the characteristics of the mammalian nose is that the tip is covered with hairless skin that is kept constantly moist; the upper lip of mammals such as dogs and cats is grooved and firmly tethered to the underlying gum. Higher primates have dry, hairy noses, and the upper lip is free. Among other things, a free upper lip gives the face a much greater range of expression than a grooved harelip.

Sense of sight: The lessening of one sense is usually associated with the intensification of another, and this is what has happened with the higher primates. Vision is acute, and objects near or distant can be brought sharply into focus. Color vision is present, as is the ability to perceive depth and distance (stereoscopic vision). This has been brought about by the change in the position of the eye sockets in the skull. In most mammals the sockets predominantly face sideways, and the fields of vision of the two eyes do not overlap each other; hence there can be no stereoscopic vision. In higher primates the eye sockets have moved around to the front of the skull, with the result that the field of vision of each eye overlaps the other. Migration of the eye sockets was in part made possi-

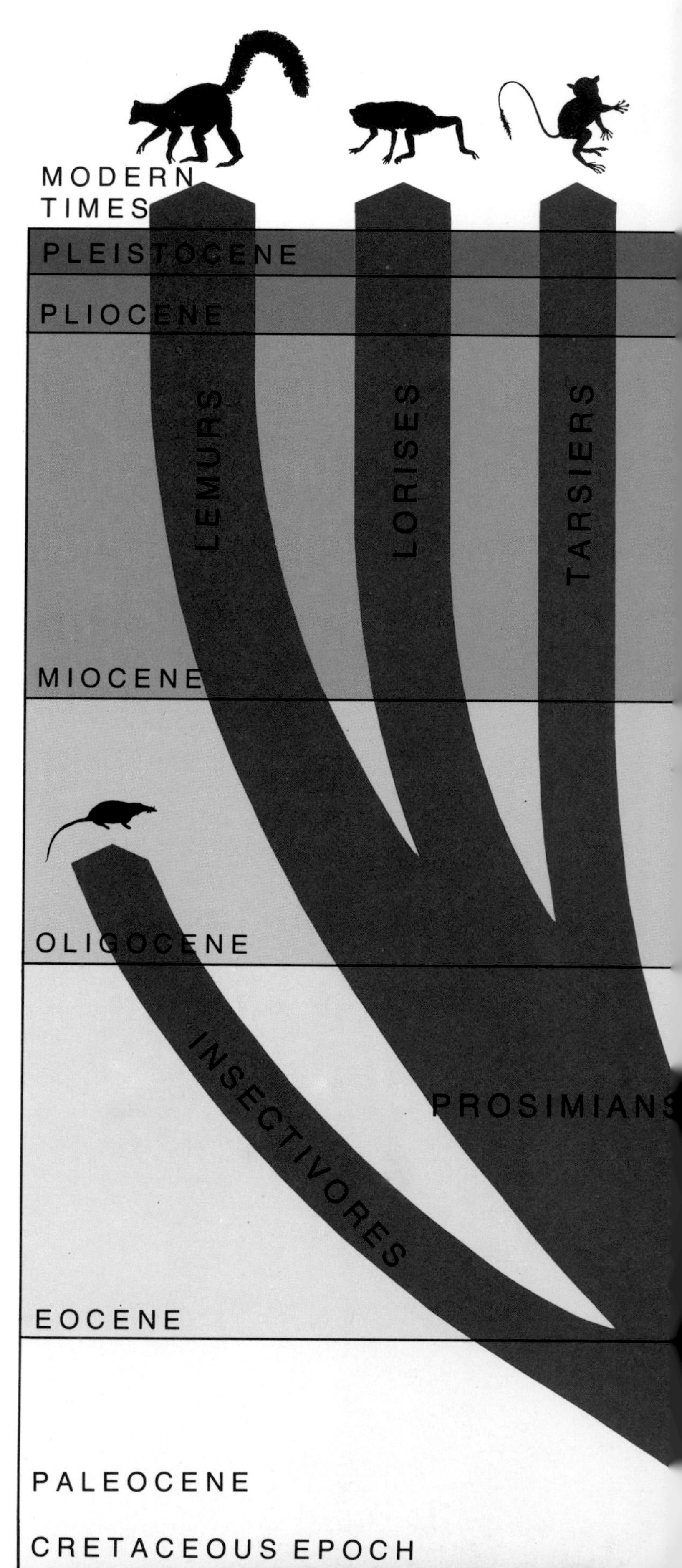

Primate evolution began with primitive, tree-dwelling insectivores and continued as successive species improved their ability to maneuver in the trees. About 60 million years ago the prosimians began to emerge while the ancestral insectivores developed long, thin fingers and the clawed animals developed the ability to grasp. This became a vital trait, for it allowed the primates to climb out of reach of predators and increased their ability to obtain food. The prosimians dominated the trees of the Americas, Europe and Asia for about 20 million years. Monkeys, however, eventually displaced prosimians as lords of the trees, and by 30 million years ago one group had traveled across the Americas (New World Monkeys) while another spread over Africa, southern Europe and southern Asia (Old World Monkeys). The apes, from whom man was to evolve, emerged between 25 and 30 million years ago. Their short, wide trunks and long, free-swinging arms distinguished them from the monkeys, and their greater size gave them a competitive edge and a longer life-span.

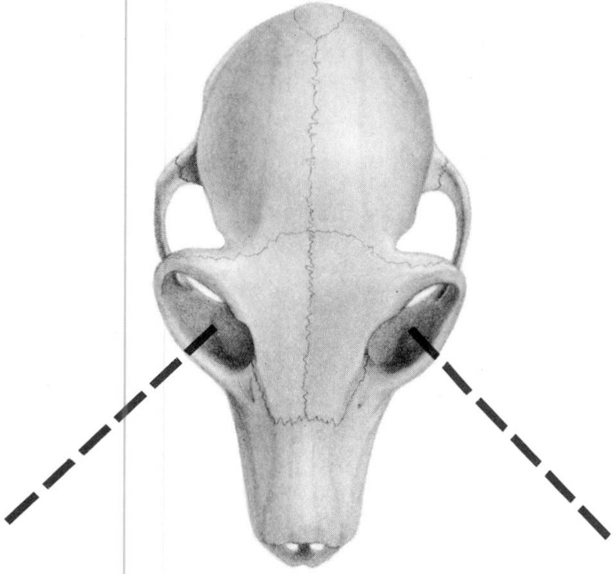

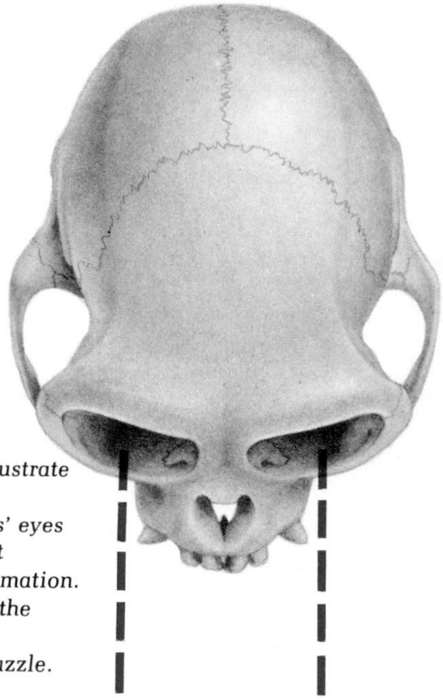

The skulls of a lemur (above, left) and a langur (above, right) illustrate the anatomical adaptations primates underwent as the relative importance of their senses changed. The more primitive lemurs' eyes face sideways, each receiving a different image. The long snout indicates a great reliance on smell for gathering food and information. The langurs' eyes face front, giving them binocular vision and the ability to judge objects more accurately. This lessened their dependence on smell and gradually resulted in a shortened muzzle.

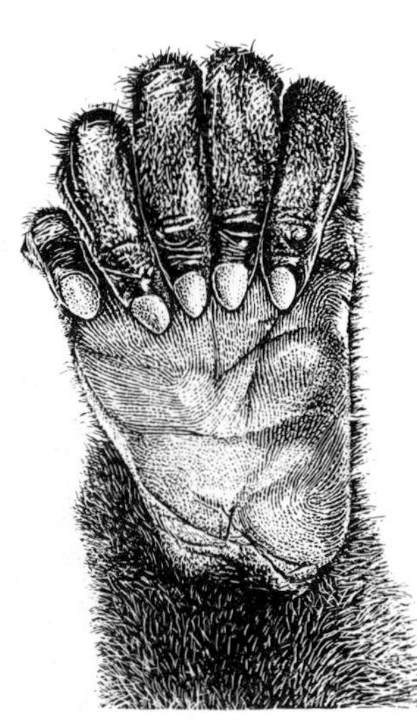

The hands of New World Monkeys (above) have curved nails and semiopposable thumbs. Old World Monkey hands (below) differ in that they have thumbs that are fully opposable, allowing these monkeys to perform precise and coordinated movements, such as plucking the thinnest blade of grass.

ble by the shortening and narrowing of the nose and a concomitant reduction of the sense of smell. The evolutionary improvement in the visual sense has led to the enlargement of the visual centers in the back of the brain (the occipital lobe), which accounts for the somewhat bulbous shape of the back of the head that is characteristic of higher primates.

Sense of hearing: In contrast to the large mobile ears of mammals such as hares that are constantly scanning the environment like a pair of radar saucers, the ears of higher primates are small, simple affairs that are practically immovable. The source of sound is located, somewhat clumsily, by turning the head from side to side. Clearly mammals that can swivel both ears toward the source of a sound have an advantage over those that cannot, but primates are amply compensated by having such excellent vision.

Sense of touch: Along with improved vision, the most important refinement of the senses in primates is in the sensitivity of the tips of the digits as organs of touch. Nonprimate mammals use their noses, their mouths and their sensitive whiskers for keeping in touch with the world about them, but primates have the additional advantage of 20 extra organs of touch—one at the end of each finger and toe. In practice the feet are usually fully involved in supporting the animal, and it is the hands that are the most valuable sensory explorers.

The basis for this important advance is the evolution of claws into nails. Claws, which grow from the ends of the digits, are long, curved and often extremely sharp, whereas nails, which are situated on the backs of the fingertips, are short and blunt. Consequently the tips of the fingers, no longer hindered by projecting claws, were free to develop as organs of touch.

Closely associated with the important sensory role of the hands is the ability to grasp. Grasping is achieved by the combined action of the fingers and thumb, which wrap over an object and stabilize it firmly against the palm, a movement that can be performed only in the absence of claws. In higher primates the thumb is set apart from the fingers and enjoys a degree of independent movement that is particularly well developed in Old World Monkeys and apes. The thumb in this group is said to show *opposition*. When the thumb is swung across the palm in the process of grasping, it undergoes a 90° rotation about its own axis, with the result that at the end of the movement the tip of the thumb lies diametrically opposite the tips of the fingers. New World Monkeys lack the element of rotation, and thus full opposability is impossible. A thumb capable of opposition makes it possible to carry out delicate manipulative tasks. The "grooming" action of Old World Monkeys who pick at one another's fur to remove dirt and flakes of body dandruff is a good example of opposition in action.

Finally, mention must be made of the most important and significant characteristic of higher primates—the ability to sit, stand and even walk upright on two legs. There are of course certain mammals that habitually move about on two legs, such as kangaroos; they hop but they do not walk. Only bears naturally walk bipedally, but, unless they are circus-trained, they seldom take more than a few steps. Certain higher primates for various reasons, such as the need for carrying food in their arms, may run or walk considerable distances bipedally. Chimpanzees, macaque monkeys, spider monkeys, gibbons and gorillas are particularly notable for their expertise as two-footed walkers, but even so they are mere amateurs when compared with bipedal man.

The special interest that scientists have in the primates is due to the fact that man is a primate too; his body is built on similar lines. This was first recognized over 2,000 years ago by a Greek physician, Galen of Pergamum, who directed his students to dissect monkeys instead of human cadavers because, as he wrote, "the ape is likest to man." The similarity goes beyond structure and extends even to diseases to which man and other primates are susceptible and other mammals are not. Thus in one sense the study of primates is the study of man's background—his evolutionary heritage.

Gorillas

The gorilla, largest of all the primates, is an awesome physical specimen. A full-grown male may weigh as much as 600 pounds, the equivalent of three football fullbacks rolled into one, and often reaches six feet in height. He has black fur, beetling brows, long arms, short legs and huge hands and feet. He walks in a quadrupedal, "knuckle-walking" manner, taking the weight of his forequarters on the backs of his knuckles. Even his head adds to his menacing look, with its high forehead and small ears set flat against the sides.

Yet despite their great size and fierce aspect, gorillas have remarkably peaceful dispositions and lead generally tranquil lives. Found only in the tropical forests and mountains of central Africa, they are strict vegetarians. Because of their size, they feed and live mainly on the ground. They consume great quantities of food, and their heavy jaws and powerful jaw muscles can handle almost any type of vegetation—bark, pith, stems and roots, as well as leaves and fruit. Much of their day is spent eating tree ferns and vines or, in mountainous regions inhabited by one subspecies, bamboo shoots and wild celery.

Gorillas live in troops of five to 30 individuals. Usually one mature silver-backed male (so named because of the saddle of white hair that grows across his back) acts as leader and protector of the group. In a typical group there might be one leader, one or two young black-backed males, six adult females and nine or ten assorted offspring. Their ordinary day is a relaxed affair. Inasmuch as they are among the most powerful animals in the jungle, gorillas have few enemies. There is no need to hunt for food; it is abundant at all seasons all over their natural habitat. They seem to know what they want to eat and where to get it. The first few hours of the day are devoted more or less exclusively to feeding. This is followed by a leisurely trek through the forest, during which little nibbles of food can be plucked in passing. At the hottest time of the day the group's activity slumps to near zero. Some gorillas build day nests for midday siestas; others simply lie down wherever they happen to be.

During the somnolent time of siesta, while the adults are dozing or just sitting around, the young indulge in games not unlike those of human children—somersaulting down a slope, hanging upside down from a branch, swinging on a vine Tarzan-style, chasing one another, wrestling or clambering over the bodies of their reclining elders. Such play develops bodily skills and allows young gorillas to learn about one another and the world they live in.

Grooming usually takes place during the siesta period. Mothers curry their babies while other adults and juveniles usually groom or scratch one another, although not with the regularity or enthusiasm shown by other primates.

Throughout the long afternoon the group moves slowly, feeding intermittently until dusk. Night falls swiftly in equatorial Africa. The silver-backed male makes his nest first and the others follow suit. Because of his enormous weight, the male's nest is almost always located on the ground and consists of a rough platform of branches and herbage bent inward from many angles. Occasionally the lighter females and young take to the trees at night, choosing a fork or crotch about 10 feet from the ground as a nest site, bending leafy branches inward to make a comfortable bed. As darkness falls, all activity ceases and the gorilla troop settles down for the night.

It is not difficult to understand why man has given the noncarnivorous gorilla its ill-deserved reputation for ferocity. One reason is certainly the gorilla's characteristic chest-beating display. The full threat ritual of an adult male gorilla is indeed chilling. First he hoots softly, then gradually breaks into a growl. Rising to his feet and hunching his shoulders, he beats his chest with cupped hands to make a hollow sound—"pok-pok-pok"—like hitting an empty gourd with a stick. Running on two feet at first, then dropping onto all fours, he tears noisily through the undergrowth, swatting at anything in his path. Finally he thumps the ground with his palms.

The function of this display is probably to release the tension that has built up in response to some disturbing stimulus and to assert his dominance over the rest of the troop. An adult male gorilla thumps his *own* chest and swipes harmlessly at bushes rather than attack another gorilla. Furthermore it seems to work: Gorillas live to be 40 or 50 years old—if unmolested by man.

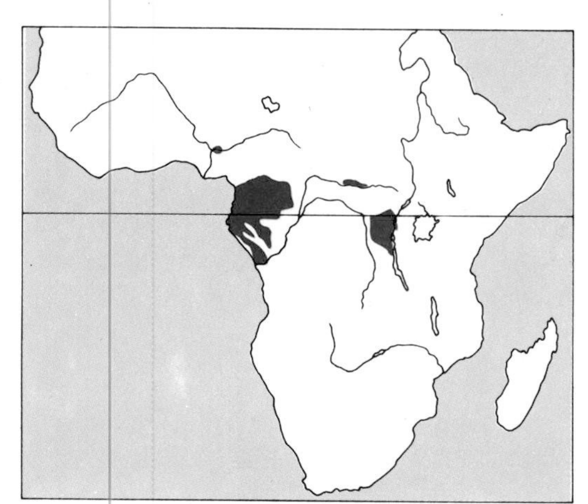

Gorillas live in the lowlands between the Niger and Congo rivers and in the mountains of central equatorial Africa.

A Way with Babies

The largest primate family is divided into three gorilla subspecies—the western lowland gorilla, which inhabits the jungles of West Africa between the Niger and Congo rivers; the eastern lowland gorilla, which ranges from east of the upper Congo to southwest Uganda; and the shaggy mountain gorilla of the volcanic highlands around Lakes Tanganyika and Albert. The mountain gorilla was not discovered until 1901 and was hunted almost to extermination before the vast Albert Park reserve was established and protective laws enacted. The three subspecies vary slightly in coloration—the lowland gorilla is thinner and has a lighter coat than its mountaineer cousin—but their habits and natures are identical, even though they live in very different, widely separated environments. In each species the female gorilla is a fond and excellent mother and the male a tolerant father. Born after a gestation of eight and one half months, young gorillas are completely dependent on their mothers for food, transportation, protection and emotional support until they reach the age of three, when they become independent members of a troop. But even in their "juvenile" years—from three to six—young gorillas will return to their mothers for a reassuring hug when they are disturbed. One orphaned gorilla, the celebrated albino Snowflake (right), captured the heart of the world when his mother was shot and, with the help of human foster parents, he became the frolicsome star of the Barcelona Zoo.

Copita de Nieve—*Little Snowflake—was born in the jungles of Rio Muni, now part of the Republic of Equatorial Africa, and was captured and brought to Barcelona when his mother was shot during a raid on a banana patch. The only albino gorilla ever known to science, he immediately attracted the worldwide interest of primatologists, and his playful antics and friendly nature endeared him to the zoogoing public. When he reached maturity, he was successfully mated to two normally colored female gorillas, begetting a son and a daughter. Though neither baby is a mutant, zoologists hope that, through cross-breeding, Snowflake will eventually sire other white gorillas.*

Contented and secure, a baby gorilla naps in its mother's arms after a midday nursing (left). At the age of eight months babies are weaned and begin to practice making their own clumsy nests each night, but they continue to sleep with their mothers until they are three and to seek their affection for another three years. Gorilla mothers are devoted parents and ferocious defenders of their young.

Gorilla females give birth to an infant about once every four years. It is often difficult to recognize a pregnant female, since their abdomens are normally very well developed. The baby gorilla is born after almost nine months and is completely dependent on its mother. It receives its sustenance almost entirely from her ample breasts until it is eight months old.

Members of a mountain gorilla troop forage for bwamba fruits in a bean field in Zaire. Except for a short siesta in midmorning, gorillas spend almost all their waking hours eating.

A young gorilla sleeps in the same nest with its mother until it is three years old. During those years the youngster has gradually learned to eat wild celery, thistles, fruit and bamboo, like the young mountain gorilla shown at left, sampling a bamboo stalk in the Kahuzi-Biega National Park, Zaire.

A young mountain gorilla climbs up a vine-covered bank with a choice bit of foliage in its mouth. A single gorilla consumes an enormous amount of greenery in a day. It takes a lot of food to sustain a fully mature body that may weigh as much as 600 pounds, including an ample potbelly.

Leader of the Troop

A troop of gorillas in the wild is always dominated by the oldest, strongest male, who makes all decisions—when to get up in the morning, where to look for food, when and where to sleep—and disciplines unruly or challenging younger members, usually with nothing more than a baleful glare. As a means of emphasizing his authority or of frightening intruders away, the leader enacts his fearful "dance," thumping his chest, racing through the underbrush, flailing away at anything in his path (below). The ritual, though, is hardly ever more than a threat: Even a charging gorilla will almost always stop short of an actual attack. In the rare instances when a gorilla has attacked a man, the encounters have invariably ended with a single bite. No gorilla has ever been known to kill a human. Other creatures, with the exception of an occasional leopard, give gorillas a wide berth.

No Room in the Ark

by Alan Moorehead

Alan Moorehead worked as a reporter in Africa during World War II and returned to that continent some years later in a different capacity, that of a naturalist. His book No Room in the Ark *represents his observations during four African expeditions and reflects his growing concern with the need for the preservation of wildlife.*

The gorilla, an awesome creature, is nowhere better described than in the short passage that follows from No Room in the Ark. *Traveling in southern Uganda, Moorehead and a friend, exhausted from hours of tramping through the bush, had the rare opportunity of meeting a gorilla.*

In the earlier part of our journey through Africa my friend and I had always chosen a pleasant place with a view or beside some stream in which to eat our lunch. Here, however, we sank down on to the earth where we were and dully stuffed the food into our mouths. There was no view anywhere, nothing but the oppressive and silent scrub. The two guides watched us impatiently, squatting a yard or two away. Yet it was amazing what those hunks of bread and meat did: life and hope began to flow through the blood again, and a cup of sweet coffee from a thermos flask accelerated the process. I rose groggily to my feet and faced the impossible once more. We fell into line again with myself drawing up in the rear.

The guides now adjured us to keep the strictest silence, and in fact it was this silence that dominated all the last moments of our climb. It closed around one with a thick palpable druglike heaviness, almost as if one's ears were stuffed with cotton wool or one's sense of hearing had suddenly failed; layer on layer of silence. And this void, this nothingness of sound, was suddenly torn apart by a single high-pitched bellowing scream. It was bizarre to the point of nightmare. It was as if one had received a sudden unexpected blow on the back of the head. As I stood there, heart thumping, transfixed with shock, one of the guides grabbed me by the arm and half dragged and half pushed me through the undergrowth towards a little rise where the others were standing. I looked at the point where they were staring and I remember calling out aloud, 'Oh my God, how wonderful!'

And the truth is he was wonderful. He was a huge shining male, half crouching, half standing, his mighty arms akimbo. I had not been prepared for the blackness of him; he was a great craggy pillar of gleaming blackness, black crew-cut hair on his head, black deep-sunken eyes glaring towards us, huge rubbery black nostrils and a black beard. He shifted his posture a little, still glaring fixedly upon us, and he had the dignity and majesty of prophets. He was the most distinguished and splendid animal I ever saw and I had only one desire at that moment: to go forward towards him, to meet him and to know him: to communicate. This experience (and I am by no means the only one to feel it in the presence of a gorilla) is utterly at variance with one's reactions to all other large wild animals in Africa. If the lion roars, if you get too close to an elephant and he fans out his ears, if the rhinoceros lowers his head and turns in your direction, you have, if you are unarmed and even sometimes if you are, just one impulse and that is to run away. The beast you feel is savage, intrinsically hostile, basically a murderer. But with the gorilla there is an instant sense of recognition. You might be badly frightened, but in the end you feel you will be able to make some gesture, utter some sound, that the animal will recognize and understand. At all events you do not have the same instinct to turn and bolt.

Afterwards I remembered another thing. Normally, when you come up against a rare wild animal in Africa, you grab your binoculars or your camera at once. It is a reflex action. This gorilla was thirty yards away and divided from us by tangled undergrowth and might not perhaps have made a very good photograph, but we could certainly have seen him more clearly through glasses. Yet none of us moved. In my own case (and I suspect in the case of my friend as well) I felt that there was not a second to be lost of this contact, not even the few instants required to put the binoculars to my eyes. I wanted to see him naturally and I wanted to see him whole.

And now abruptly he rose to his full height. Had I really been about to give expression to my sub-conscious desire to move towards him I expect that, at this moment, I would have paused, for he was tremendous in his great height and strength. It was a question now as to whether or not he would beat on his chest and charge, so as to give his family (unseen by us but certainly lurking somewhere there in the bush) further time to get away, but, in fact, he did neither of these things. He lifted his head and gave vent to another of those outlandish and terrifying barking-screams. Once again it seemed to bring every living thing in the bush, including one's own heart, to a full stop. Then he dropped on to his hands and melted away. There was, of course, no chance of following him; despite his size he could travel many times faster than we could.

That was the end of the show, and it had lasted I suppose a couple of hundred seconds. Yet still, after much wandering through Africa in the last few years, I rate this as the most exciting encounter that has come my way; and I remember how, no longer any need for silence, the guides with their pangas slashed a path for us to return through the bush, and how they grinned and were pleased because we were pleased, and how I went down the mountain like a young gazelle in two hours straight, never a touch of fatigue, never a thought for my blistered feet after such a happy day.

The Year of the Gorilla *by George B. Schaller*

In 1959 George Schaller, a pioneer in the field of ethology, set out on an expedition to study the mountain gorilla of the western Congo region. Although much had been written about the gorilla over the years, little concrete information had been obtained. Hunters tended to stress the seeming ferocity of the gorilla, while early naturalists were unable to get more than a fleeting impression of the animal. Schaller and his wife, Kay, approached the problem in a new way. For more than a year they lived among the gorillas, learning to know them as individuals, studying their changing habits and moods and sharing their world. The results of Schaller's study have given scientists new insights into primate behavior. The following portraits from The Year of the Gorilla *give the reader a true impression of the gorilla and lay to rest the image of the animal as belligerent.*

I followed the trails of the gorillas through the trampled vegetation, never certain if the animals had gone a hundred yards, a mile, or doubled back so that now they were somewhere behind me. The trails always had an interesting story to tell, and in many ways I enjoyed my saunterings as much as the sight of the apes themselves. Almost unconsciously I fell into the same unhurried pattern of movement as that of the gorillas, especially when the morning sun shimmered on the leaves and the mountains reached serenely into the sky. When gorillas were feeding, they fanned out, leaving many trails littered with discarded celery bark and other food remnants. When the gorillas were traveling, they moved in single or double file, only to rest after awhile close together on an open slope. Sometimes a musty odor, like that of a barnyard, permeated the air, and I knew that it was the site where the animals had slept the night before. It often took me over half an hour to find all the nests, since a gorilla occasionally slept sixty feet or more from its nearest neighbor. I mapped each site and paced off the distance between nests. By measuring the diameter of dung in the nests, I could determine the resting place of large silverbacked males and of juveniles. Frequently both medium and small-sized sections of dung lay side by side in the same nest, indicating that a female and her offspring had slept in it. . . .

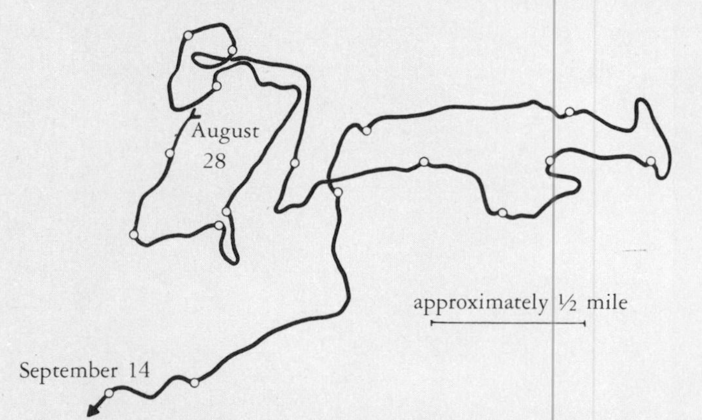

The daily route of travel of a group of gorillas on the slopes of Mt. Mikeno between August 28 and September 14, 1959, as recorded by George Schaller. Each circle represents one night's nest site.

Animals are better observers and far more accurate interpreters of gestures than man. I felt certain that if I moved around calmly and alone near the gorillas, obviously without dangerous intent toward them, they would soon realize that I was harmless. It is really not easy for man to shed all his arrogance and aggressiveness before an animal, to approach it in utter humility with the knowledge of being in many ways inferior. Casual actions are often sufficient to alert the gorillas and to make them uneasy. For example, I believe that even the possession of a firearm is sufficient to imbue one's behavior with a certain unconscious aggressiveness, a feeling of being superior, which an animal can detect. When meeting a gorilla face to face, I reasoned, an attack would be more likely if I carried a gun than if I simply showed my apprehension and uncertainty. Among some creatures—the dog, rhesus monkey, gorilla, and man—a direct unwavering stare is a form of threat. Even while watching gorillas from a distance I had to be careful not to look at them too long

without averting my head, for they became uneasy under my steady scrutiny. Similarly they considered the unblinking stare of binoculars and cameras as a threat, and I had to use these instruments sparingly. . . .

Early in the study I had noted that the gorillas tend to have an extremely placid nature which is not easily aroused to excitement. They give the impression of being stoic and reserved, of being introverted. Their expression is usually one of repose, even in situations which to me would have been disturbing. All their emotions are in their eyes, which are a soft, dark brown. The eyes have a language of their own, being subtle and silent mirrors of the mind, revealing constantly changing patterns of emotion that in no other visible way affect the expression of the animal. I could see hesitation and uneasiness, curiosity and boldness and annoyance. Sometimes, when I met a gorilla face to face, the expression in its eyes more than anything else told me of his feelings and helped me decide my course of action. . . .

When I began to study gorillas, I was at first tremendously impressed by their human appearance—they gave the superficial impression of slightly retarded persons with rather short legs wrapped in fur coats. These gestures and body positions of gorillas, and for that matter also those of other apes, resemble those of man rather than the monkeys. They stretch their arms to the side and yawn in the morning when they wake up, they sit on a branch with legs dangling down, and they rest on their back with their arms under the head. The great structural similarity between man and apes has been noted repeatedly since the time of Linnaeus and Darwin, and it is for this reason that all have been placed taxonomically into the super-family *Hominoidea*. In their emotional expressions too the gorillas resemble man: they frown when annoyed, bite their lips when uncertain, and youngsters have temper tantrums when thwarted. The social interactions between members of a gorilla group are close and affectionate, much like that of a human family, and their mating system is polygamous, a type for which man certainly has a predilection. These and many other basic similarities are to be expected, for man and the apes evolved from a common ancestral stock of monkey-like apes which diverged, one line leading to the apes, the other to man. It must be assumed from the evidence of evolution that man became man by the slow accumulation of certain characteristics, he became man by degrees, but still retained in his mind and frame the stamp of his origin. . . .

Chimpanzees

The chimpanzee is man's closest relative. Fundamental similarities show that the two evolved from a common ancestral stock, although exactly when the two stocks separated is still hotly debated. Some scientists say that it was not less than 20 million years ago; others, not more than five million.

The chimpanzee's black coat has roughly the same number of hairs as a man's body, although each of its hairs is longer, darker and thicker and so more conspicuous. Its torso is virtually identical to man's except in its proportions. Long arms and short legs give the chimp a lumbering quadrupedal gait, similar to that of the gorilla and very different from man's upright, striding walk. Weighing about 150 pounds at maturity, a male chimp grows to about four and a half feet tall.

Even in the reproduction process chimpanzees and human beings are similar in many respects. The menstrual cycle in chimpanzees lasts about 35 days, with ovulation occurring midway in the cycle. The main difference is that in the female chimpanzee ovulation coincides with a week to ten-day period when she is sexually receptive to the male, a condition known as estrus; in contrast, the human female is receptive throughout her cycle. Estrus in chimpanzees is accompanied by a large pink swelling on the female's hindquarters that indicates her readiness to mate. There is little competition among the males for females. All the males in a chimpanzee group will attempt to mate with each female; thus the father of an infant can never be identified.

The newborn chimp, weighing about four pounds, is almost as weak and helpless as a human baby. Cradled in its mother's arms, it can reach her nipples and feeds at least once an hour. Soon the young infant learns to ride beneath its mother's belly, gripping her fur with hands and feet. By the age of six months it is riding on her back as she moves through the forest. At this stage the infant has an appealing white tail tuft, the mark of its privileged status as a baby who must be treated kindly by every member of the group. But as the white tuft gradually disappears, the juvenile's status subtly changes, and the youngster is subject to discipline if it transgresses the rules of chimpanzee society.

Chimpanzees live in and around the tropical rain forests of Africa from Sierra Leone to Tanzania, where their basic diet of fruit, nuts and young leaves—with occasional feasts of meat—is available all year round. Social animals above all, they live in groups of 30 to 80 individuals that are constantly in the process of changing composition. There is little rivalry among adult males, and the dominant male system governing other primate troops is less apparent among the sociable chimps. To exploit the scattered food resources of the forest the group splits up into small bands. Five or six adult males, three or four mothers and their young, or mixed parties of males and females, adults and young, travel their range in search of food. They keep in touch by calling to one another. When one band finds a rich food source, their excited barks alert other bands in the neighborhood.

Chimpanzees are capable of a wide range of activities. They hunt, build sleeping nests, use stones to crack nuts and sticks to investigate holes in trees. They use and make tools. They "fish" for termites by inserting a piece of stick into the holes of the nest and carefully withdrawing it covered with clinging termites. In order to make a "fishing rod," a chimpanzee selects a twig and improves it by removing any projections.

After a chance hunting success chimps will sometimes acquire a taste for meat and actively go after prey. When a young baboon or colobus monkey strays away from its troop, chimpanzees may seize and kill it. The carcass is then dragged up a tree, where the owner is besieged by other chimpanzees, holding out their hands palm upward in the typical chimpanzee begging gesture. This is something that an adult male chimpanzee finds very hard to resist; he tears off fragments of meat from the carcass and hands them out. Food sharing among chimpanzees seems to be the seed of what in man has developed into one of his most esteemed qualities—unselfishness.

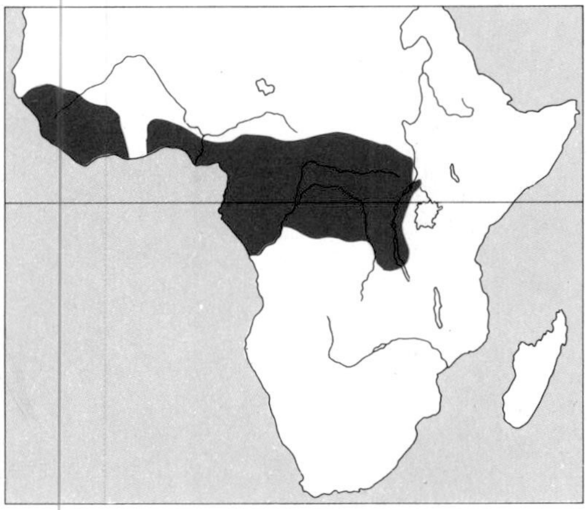

Chimpanzees are found in the rain forests of equatorial Africa from the Atlantic east to Tanzania.

The Chimps' Arboreal Abode

Chimpanzees are found only in Africa. They are equally at home in the tropical rain forests (left) and woodland savannas (below) that border the Equator. When seen from above, the tropical forest is a thick canopy of green, not easily penetrated, even by the unrelenting equatorial sun. On the forest floor twilight prevails, shedding a diffuse light on a life that is relatively easy for most primates, including the chimps. Living so close to the Equator, where there is daylight for 12 out of every 24 hours, the chimps have ample time for their daily foraging and feeding. And because there are no extreme seasonal changes in the forests and savannas, the supply of fruit, leaves, flowers and insects is always plentiful, assuring the chimpanzees food throughout the year.

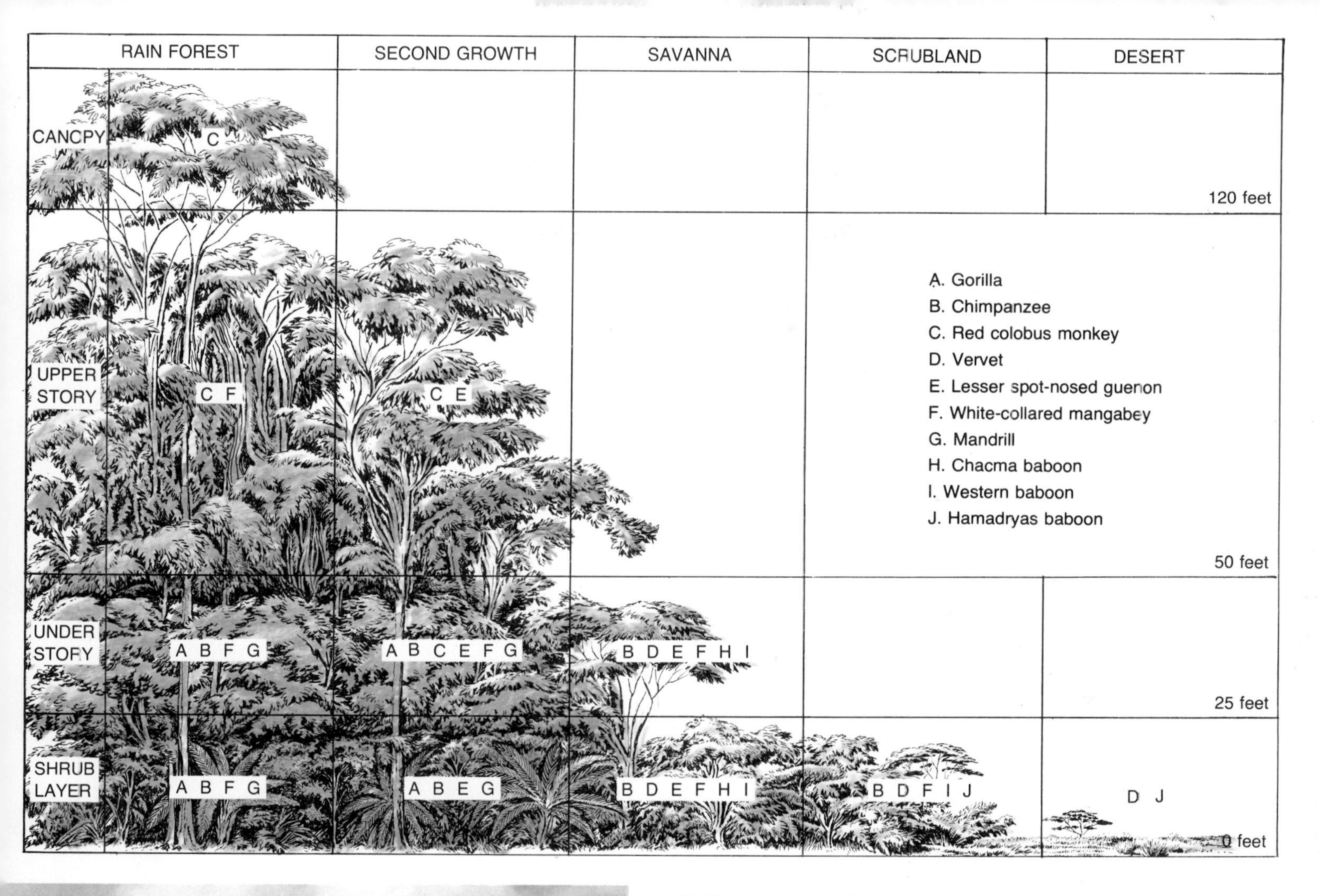

Unlike the New World Monkeys of the Western Hemisphere, almost all of which live in the uppermost branches of the forests and jungles they inhabit, the monkeys and apes of Africa live in a wide variety of habitats, from steaming rain forests to parched deserts (above). Most, however, make their home in the branches of the different strata of the rain forests. Here they are safe from predators and can always find the plants and insects they feed on. The lighter-bodied, more acrobatic monkeys, such as the red colobus monkey, the lesser spot-nosed guenon and the white-collared mangabey, inhabit the highest levels, which are often 150 feet above ground. The less agile and heavier simians, such as the chimpanzee, gorilla and mandrill, occupy the lower levels and the forest floor, where the thicker branches are capable of holding their weight. The savanna- and desert-living vervets and baboons spend many hours each day on the ground, digging up the tubers and foraging for the grasses and seeds that make up part of their varied diet.

Tender Loving Care

Chimpanzee mothers give birth to a single infant once every three to five years, which makes the arrival of a baby a rather special event. The rest of the chimp community becomes greatly excited and curious about the new addition, with males and females alike wanting to hold and examine it. After it is a few days old, the baby is able to grip its mother's fur securely enough to hang on underneath her as she forages for food or even flees from impending danger. By the time it is five or six months old the baby is strong enough to ride on its mother's back.

Although the community as a whole, including the male members, is usually tolerant of and affectionate toward its youngest members, tempers do flare, and a laughter-filled tickling session can quickly turn into an aggressive display by an adult male toward the youngster. It is therefore up to a mother to see that her child doesn't overstep the bounds of chimp etiquette. In fact, chimpanzee young depend on their mothers for several years. They share their mother's nest until they are about four years old and often suckle until then as well. By this time a young chimp has learned to manipulate objects easily and to move through its environment knowledgeably, preparing it for the imminent experience of weaning and for its future role as an adult member of the community.

For the first few months of its life the infant chimpanzee's world revolves primarily around its mother. It is from her that it learns the basic principles of chimp social life, such as grooming, which not only keeps the apes clean but also helps build friendly bonds among community members. The female on the opposite page does the familiar parental back-of-the-neck check as another older sibling relaxedly waits its turn in the tree nest the three share. The infant above joyously submits to some maternal tickling. Minor cuts always get special attention, but the youngster and its mother at left seem positively spellbound by what it has found on its foot.

TARZAN OF THE APES

by Edgar Rice Burroughs

Edgar Rice Burroughs' creation, Tarzan, has been a hero to countless readers and moviegoers in the 24 books and 40 movies that recount his adventures. Lost in the forest, John Clayton, Lord Greystoke, and his wife, Lady Alice, survived only a year after their son was born. The infant, living through one of the most unusual childhoods recorded in literature, grew up to become Tarzan, Lord of the Jungle.

In this excerpt from Tarzan of the Apes *we are first introduced to Kala, the fiercely protective she-ape who became Tarzan's surrogate mother, and learn the circumstances of Tarzan's adoption. Although author Burroughs heavily anthropomorphized his animal characters, it is interesting to note the parallel behavior exhibited by Kala in this excerpt and by Olly in Jane van Lawick-Goodall's real-life account of a mother ape faced with the death of her offspring (page 40).*

In the forest of the table-land a mile back from the ocean old Kerchak the Ape was on a rampage of rage among his people.

The younger and lighter members of his tribe scampered to the higher branches of the great trees to escape his wrath; risking their lives upon branches that scarce supported their weight rather than face old Kerchak in one of his fits of uncontrolled anger.

The other males scattered in all directions, but not before the infuriated brute had felt the vertebra of one snap between his great, foaming jaws.

A luckless young female slipped from an insecure hold upon a high branch and came crashing to the ground almost at Kerchak's feet.

With a wild scream he was upon her, tearing a great piece from her side with his mighty teeth, and striking her viciously upon her head and shoulders with a broken tree limb until her skull was crushed to a jelly.

And then he spied Kala, who, returning from a search for food with her young babe, was ignorant of the state of the mighty male's temper until suddenly the shrill warning of her fellows caused her to scamper madly for safety.

But Kerchak was close upon her, so close that he had almost grasped her ankle had she not made a furious leap far into space from one tree to another—a perilous chance which apes seldom if ever take, unless so closely pursued by danger that there is no alternative.

She made the leap successfully, but as she grasped the limb of the further tree the sudden jar loosened the hold of the tiny babe where it clung frantically to her neck, and she saw the little thing hurled, turning and twisting, to the ground thirty feet below.

With a low cry of dismay Kala rushed headlong to its side, thoughtless now of the danger from Kerchak; but when she gathered the wee, mangled form to her bosom life had left it.

With low moans, she sat cuddling the body to her; nor did Kerchak attempt to molest her. With the death of the babe his fit of demoniacal rage passed as suddenly as it had seized him.

Kerchak was a huge king ape, weighing perhaps three hundred and fifty pounds. His forehead was extremely low and receding, his eyes bloodshot, small and close set to his coarse, flat nose; his ears large and thin, but smaller than most of his kind.

His awful temper and his mighty strength made him supreme among the little tribe into which he had been born some twenty years before.

Now that he was in his prime, there was no simian in all the mighty forest through which he roved that dared contest his right to rule, nor did the other and larger animals molest him.

Old Tantor, the elephant, alone of all the wild savage life, feared him not—and he alone did Kerchak fear. When Tantor trumpeted, the great ape scurried with his fellows high among the trees of the second terrace.

The tribe of anthropoids over which Kerchak ruled with an iron hand and bared fangs, numbered some six or eight families, each family consisting of an adult male with his females and their young, numbering in all some sixty or seventy apes.

Kala was the youngest mate of a male called Tublat, meaning broken nose, and the child she had seen dashed to death was her first; for she was but nine or ten years old.

Notwithstanding her youth, she was large and powerful—a splendid, clean-limbed animal, with a round, high forehead, which denoted more intelligence than most of her kind possessed. So, also, she had a great capacity for mother love and mother sorrow.

But she was still an ape, a huge, fierce, terrible beast of a species closely allied to the gorilla, yet more intelligent; which, with the strength of their cousin, made her kind the most fearsome of those awe-inspiring progenitors of man.

When the tribe saw that Kerchak's rage had ceased they came slowly down from their arboreal retreats and pursued again the various occupations which he had interrupted.

The young played and frolicked about among the trees and bushes. Some of the adults lay prone upon the soft mat of dead and decaying vegetation which covered the ground, while others turned over pieces of fallen branches and clods of earth in search of the small bugs and reptiles which formed a part of their food.

Others, again, searched the surrounding trees for fruit, nuts, small birds, and eggs.

They had passed an hour or so thus when Kerchak called them together, and, with a word of command to them to follow him, set off toward the sea.

They traveled for the most part upon the ground, where it was open, following the path of the great elephants whose comings and goings break the only roads through those tangled mazes of bush, vine, creeper, and tree. When they walked it was with a rolling, awkward motion, placing the knuckles of their closed hands upon the ground and swinging their ungainly bodies forward.

But when the way was through the lower trees they moved more swiftly, swinging from branch to branch with the agility of their smaller cousins, the monkeys. And all the way Kala carried her little dead baby hugged closely to her breast.

It was shortly after noon when they reached a ridge overlooking the beach where below them lay the tiny cottage which was Kerchak's goal.

He had seen many of his kind go to their deaths before

the loud noise made by the little black stick in the hands of the strange white ape who lived in that wonderful lair, and Kerchak had made up his brute mind to own that death-dealing contrivance, and to explore the interior of the mysterious den.

He wanted, very, very much, to feel his teeth sink into the neck of the queer animal that he had learned to hate and fear, and because of this, he came often with his tribe to reconnoiter, waiting for a time when the white ape should be off his guard.

Of late they had quit attacking, or even showing themselves; for every time they had done so in the past the little stick had roared out its terrible message of death to some member of the tribe.

Today there was no sign of the man about, and from where they watched they could see that the cabin door was open. Slowly, cautiously, and noiselessly they crept through the jungle toward the little cabin.

There were no growls, no fierce screams of rage—the little black stick had taught them to come quietly lest they awaken it.

On, on they came until Kerchak himself slunk stealthily to the very door and peered within. Behind him were two males, and then Kala, closely straining the little dead form to her breast.

Inside the den they saw the strange white ape lying half across a table, his head buried in his arms; and on the bed lay a figure covered by a sailcloth, while from a tiny rustic cradle came the plaintive wailing of a babe.

Noiselessly Kerchak entered, crouching for the charge; and then John Clayton rose with a sudden start and faced them.

The sight that met his eyes must have frozen him with horror, for there, within the door, stood three great bull apes, while behind them crowded many more; how many he never knew, for his revolvers were hanging on the far wall beside his rifle and Kerchak was charging.

When the king ape released the limp form which had been John Clayton, Lord Greystoke, he turned his attention toward the little cradle; but Kala was there before him, and when he would have grasped the child she snatched it herself, and before he could intercept her she had bolted through the door and taken refuge in a high tree.

As she took up the little live baby of Alice Clayton she dropped the dead body of her own into the empty cradle; for the wail of the living had answered the call of universal motherhood within her wild breast which the dead could not still.

High up among the branches of a mighty tree she hugged the shrieking infant to her bosom, and soon the instinct that was as dominant in this fierce female as it had been in the breast of his tender and beautiful mother—the instinct of mother love—reached out to the tiny man-child's half-formed understanding, and he became quiet.

Then hunger closed the gap between them, and the son of an English lord and an English lady nursed at the breast of Kala, the great ape.

In the meantime the beasts within the cabin were warily examining the contents of this strange lair.

Once satisfied that Clayton was dead, Kerchak turned his attention to the thing which lay upon the bed, covered by a piece of sailcloth.

Gingerly he lifted one corner of the shroud, but when he saw the body of the woman beneath he tore the cloth roughly from her form and seized the still, white throat in his huge, hairy hands.

A moment he let his fingers sink deep into the cold flesh, and then, realizing that she was already dead, he turned from her, to examine the contents of the room; nor did he again molest the body of either Lady Alice or Sir John. . . .

Kala had not once come to earth with her little adopted babe, but now Kerchak called to her to descend with the rest, and as there was no note of anger in his voice she dropped lightly from branch to branch and joined the others on their homeward march.

Those of the apes who attempted to examine Kala's strange baby were repulsed with bared fangs and low menacing growls, accompanied by words of warning from Kala.

When they assured her that they meant the child no harm she permitted them to come close, but would not allow them to touch her charge.

It was as though she knew that her baby was frail and delicate and feared lest the rough hands of her fellows might injure the little thing.

Another thing she did, and which made traveling an onerous trial for her. Remembering the death of her own little one, she clung desperately to the new babe, with one hand, whenever they were upon the march.

The other young rode upon their mothers' backs; their little arms tightly clasping the hairy necks before them, while their legs were locked beneath their mothers' armpits.

Not so with Kala; she held the small form of the little Lord Greystoke tightly to her breast, where the dainty hands clutched the long black hair which covered that portion of her body. She had seen one child fall from her back to a terrible death, and she would take no further chances with this.

Food for Thought

Whether they live in the dense and humid rain forest or in the drier woodland savannas, chimpanzees spend between six and eight hours a day foraging and feeding. Chimps eat communally, and each member knows its place in the hierarchy. The top-ranking male is usually the first to eat, followed by lesser males, the females and the young. Chimpanzees are omnivorous, and their diet includes leaves, vines, shoots, eggs, fruits, insects and, on occasion, meat. The primary food for the savanna chimp is grapefruit, and the average adult polishes off a total of about 16 pounds of them each day. The real gourmets among them will remove the fruit's white inner membrane, while the most finicky, when they eat grapefruit, just suck out the juice. Because of the open spaces in the savannas, these apes often have to move their food over great distances to find safe eating spots. Although their only predators are man and the leopard, the naturally shy chimpanzees prefer eating in privacy.

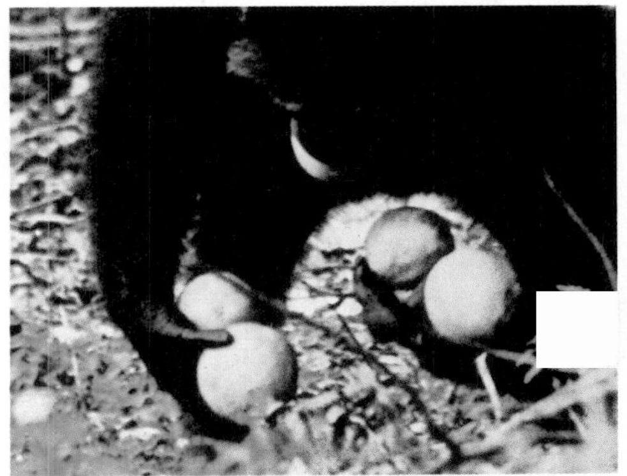

The chimpanzees on these pages are members of a forest-dwelling troop and a savanna troop being observed by a team of behavorial scientists. Bananas do not grow naturally in the forest home of the chimp on the opposite page. But when the chimp sees the fruit supplied by the scientists, it knows instinctively that the banana has to be peeled. The savanna chimp (above), in an admirable display of coordination, moves its cache of grapefruit to a more secluded spot. At right, from the top, a chimp finds a pile of grapefruits and picks up as many as it can carry, even toting one in its mouth. Each chimp has its own peeling technique, using any imaginable combination of hands, feet, teeth and lips to complete the job.

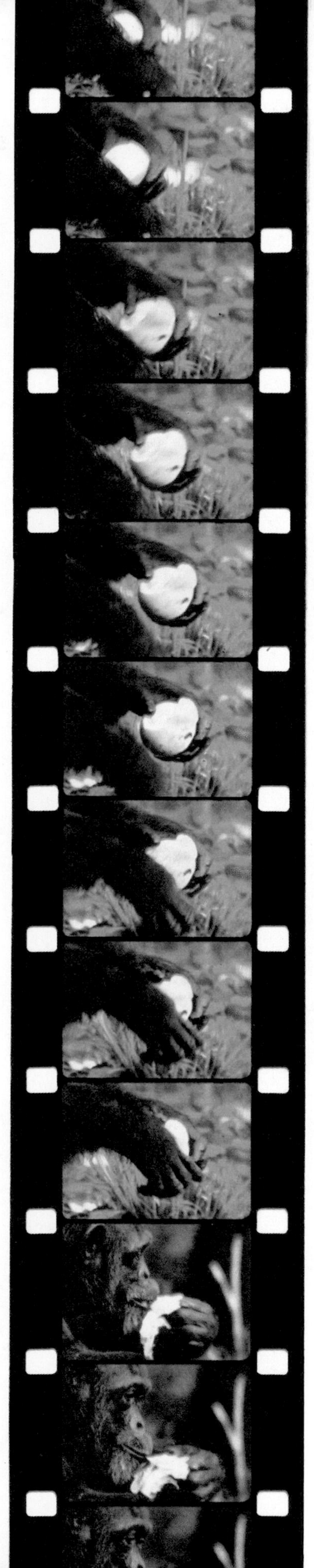

The savanna chimps on these pages were studied by a team of Dutch scientists and filmed in the wild. The chimpanzees eat at regular hours twice a day—breakfast in the morning and dinner late in the afternoon. The food is the same at both meals. Grapefruits are often picked off the lower branches of trees by apes standing on their hind legs (left). As the filmstrip on the opposite page shows, the chimps demonstrate great dexterity in peeling the fruit, even removing the white inner membrane in some instances. The animals above round out their diet with (top) termites fished out of the ground with a twig, (center) leaves that are chewed into a gumlike substance that seems to go well with insects and (bottom) other fruits like figs and dates.

In the Shadow of Man

by Jane van Lawick-Goodall

Over a period of more than ten years, Jane van Lawick-Goodall, a young Englishwoman, devoted herself to the first sustained study of chimpanzees in their natural habitat. Laboratory experiments had shown that the chimpanzee is man's closest relative, but because of the difficulties of field study, there was still much to be learned. Mrs. van Lawick-Goodall spent almost four years in the Gombe preserve in Tanzania before she was able to establish enough rapport with the chimpanzees to allow close scrutiny of their behavior. Her diligence was well rewarded; in the years that followed, the chimpanzee community accepted her and allowed her free rein for her study.

The following excerpt from In the Shadow of Man*, the story of the death of an infant chimpanzee, provides an interesting comparison with the story of Tarzan's adoption.*

Olly's new baby was four weeks old when he suddenly became ill. I had been excited when I heard of his birth: would his elder sister Gilka show the same fascination for him as Fifi had for Flint? And how would Olly react if she did? Though I was not at the Gombe when the baby was born, I was there a month later when one evening Olly walked slowly into camp supporting him with one hand. Each time she made a sudden movement, he uttered a loud squawk as though in pain, and he was gripping badly. First one hand or foot and then another slipped from Olly's hair and dangled down.

While Olly sat eating her bananas, Gilka groomed her mother. Often I had watched Gilka working her way closer to her small sibling's hands just as Fifi had done two years earlier when Flint was tiny. This time, however, Olly permitted her daughter to groom the baby's head and back without attempting, as she usually did, to push Gilka's hands away.

Next morning it was obvious that the baby was very ill. All his four limbs hung limply down and he screamed almost every time his mother took a step. When Olly sat down, carefully arranging his legs so as not to crush them, Gilka went over and sat close to her mother and stared at the infant; but she did not attempt to touch him then.

Olly ate two bananas and then set off along the valley, with Gilka and me following. Olly only moved a few yards at a time, and then, as though worried by the screams of her infant, sat down to cradle him close. When he quieted, she moved again, but he instantly began to call out so that once more she sat to comfort him. After traveling about a hundred yards, which took her just over half an hour, Olly climbed into a tree. Again she carefully arranged her baby's limp arms and legs on her lap as she sat down. Gilka, who had followed her mother, stared again at her small sibling, and then mother and daughter began to groom each other. The baby stopped screaming and, apart from occasionally grooming his head briefly, Olly paid him no further attention.

When we had been there some fifteen minutes it began to pour, a blinding deluge that almost obscured the chimps from my sight. During the storm, which went on for thirty minutes, the baby must either have died or lost consciousness. When Olly left the tree afterward he made no sound and his head lolled back as limply as his arms and legs.

I was amazed at the sudden and complete change in Olly's handling of her baby. I had watched a young and inexperienced mother carrying her dead baby, and even the day after its death she had held the body as though it were still alive, cradling it against her breast. But Olly climbed down the tree with her infant carelessly in one hand, and when she reached the ground she flung the limp body over her shoulder. It was as though she knew he was dead. Perhaps it was because he no longer moved or cried that her maternal instincts were no longer roused.

The following day Olly arrived in camp, followed by Gilka, with the corpse of her infant slung over her shoulder. When she sat down the body sometimes dropped heavily to the ground. Occasionally Olly pushed it into her groin as she sat; when she stood she held it by an arm or even a leg. It was gruesome to watch, and several of the young female chimpanzees went over and stared. So also did a number of baboons. Olly ignored them all.

Finally Olly wandered away from camp and she and Gilka, with me following, went some way up the opposite mountain slope. Olly seemed dazed; she looked neither to right nor to left but plodded up the narrow trail through the forest, the body slung over her neck, until she reached a place halfway up the mountainside. Then she sat down. The dead infant slumped to the ground beside her, and other than to glance down briefly Olly ignored it. She just sat staring into space, hardly moving for the next half hour save to hit away the fast-gathering swarm of flies.

Now, at last, came Gilka's opportunity to play with her sibling. It was not easy to watch. Already the corpse had begun to smell; the face and belly showed a definite greenish tinge, and the eyes, which were wide open, stared glassily ahead. Inch by inch, glancing repeatedly at her mother's face, Gilka pulled the body toward her. Carefully she groomed it, and then she even tried to play, pulling one dead hand into the ticklish spot between her collarbone and neck, and actually showing the vestige of a play-face. We had been so glad for Gilka's sake when old Olly had given birth again—but it looked as if Gilka was always to be ill-fated. Presently, with another quick glance toward her mother, Gilka carefully lifted the dead body of her sibling and pressed it to her breast. Only then did Olly's lethargy leave her for a moment. She snatched the body away, and then once more let it fall to the ground.

When the old female moved on, it was merely to plod down the same track and go back to camp. She ate two bananas, sat staring into space, and wandered off up the valley.

For another three hours I managed to keep up with the little family. Every ten minutes or so Olly just sat or lay, and Gilka, as before, groomed or tried to play with the dead body of her sibling. Eventually Olly became worried by my presence and started to walk fast, glancing at me over her shoulder. She headed into a dense thicket, and though I continued after her a brief distance, I decided I should desist. In fact I was glad to be out of the tangled vines, since in the hot and humid air the smell of death was heavy where Olly had passed, and each twig had its own quota of bloated flies.

The following afternoon Olly and Gilka arrived in camp without the body. Somewhere in the valley Olly must have finally abandoned it.

Orangutans

The orangutan is one of two Asian species of apes (the other is the gibbon) and has a surprisingly small range—less than 50,000 square miles of tropical rain forest on the islands of Borneo and Sumatra. Its physical characteristics are more exaggerated than those of the gorilla, its African counterpart. Its arms are longer, its legs shorter, and its hands and feet, with long, curved fingers and toes, are virtually large hooks, ideal for grasping fruit or holding onto the branches of trees but not of much use for anything else. The thumb is so small that its manipulation is extremely clumsy; the big toe is equally diminutive and in some cases does not even grow a nail.

The adult male orangutan is much smaller than the gorilla. He weighs only half as much—about 200 to 300 pounds—and he grows to only four to four and a half feet. Nevertheless, he cuts an arresting figure, with an arm span of seven to eight feet, a coat of rust-colored hair hanging like Spanish moss, a face doubled in size by a pair of fatty cheek pads, an inflatable throat sac and a conspicuous paunch. As if all that were not enough to attract attention, the male orang emits a series of groans "like an old man in very great pain," as one observer has described it. In the process, the throat pouch is inflated like a balloon, making deep pumping noises that build up into loud throaty roars. Large air spaces beneath the skin of the chest help to increase the resonant tone of the roar; it can be heard for well over half a mile.

Orangutans are basically solitary animals, the most common social group being the female with one or two offspring. They appear to have no special greeting gestures, and, when females meet, they studiously ignore one another. Each fully mature adult male is the patriarch of a territorial domain that encompasses the ranges of several females. His loud calls seem to act as a warning system, discouraging the approach of other males. Neighboring patriarchs rarely meet, but if by chance they do, tremendous branch-shaking displays continue until one of them retreats.

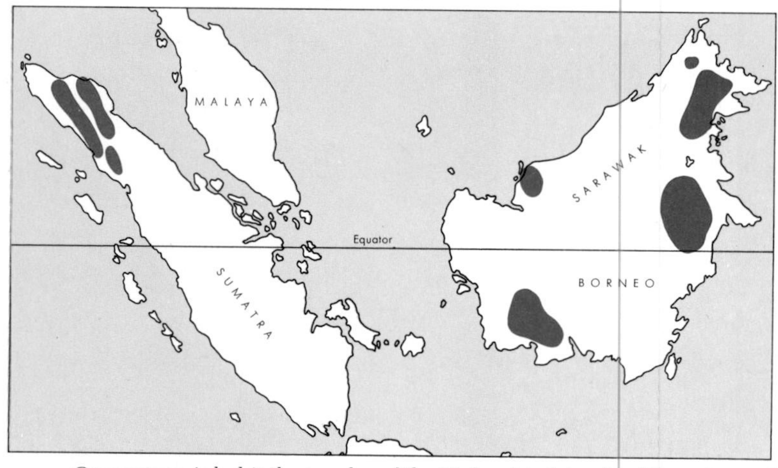

Orangutans inhabit the jungles of the Malaysian islands of Sumatra and Borneo.

Orangs, like gorillas, are vegetarians. They relish mangoes, plums, litchis, figs and, above all, the prickly durian fruit, armored like a hedgehog but well worth the trouble of opening for its delicious flesh and sweet nutty seeds. The orang's powerful hands and big shovel-shaped front teeth are invaluable for breaking open tough-shelled fruit. The incredible strength and mobility of its limbs enable it to reach fruit that would otherwise be inaccessible. A large male who cannot trust his great bulk to the smaller fruit-bearing branches of a tree simply settles himself in the middle of the tree and then bends or breaks the branches inward toward himself, quickly stripping the tree of fruit and leaving it limp and mutilated. While females and infants rarely leave the trees, the much bulkier males have occasionally been seen on the ground.

Life in the tropical forests of Indonesia is not as easy as it might seem, particularly for a fruit-eating ape that needs a large daily supply of food. Some trees bear fruit regularly, but others bear only once in every two or three years or even less. That makes the knowledge of the old patriarch invaluable, as only he knows from long experience when each tree is likely to be in fruit.

In Borneo, with fairly well-defined seasons, the summer is a time of plenty for the orangs, and they put on a lot of weight then, storing fat on their bodies as provender for the lean monsoon months from December to February, when leaves, bark and pith are their only source of nourishment.

Orangutans are the most endangered of all the apes. Their numbers are declining alarmingly in the wild, mainly because of the destruction of the forests. Poaching also takes its toll; a mother is shot and her baby is captured for export to foreign zoos or collectors. The present-day population has been estimated at fewer than 5,000 animals, and any further shrinkage might bring about their extinction in the wild.

The Ape That Relaxes Like Man

The young orangutans shown on these pages were all photographed in two forest preserves in the Malaysian states of Sabah and Sarawak, North Borneo, where an experiment aimed at reintroducing orphaned babies to their natural habitat has been going on since 1961. Many orphans are retrieved from poachers who intended them for foreign zoos and are being acclimated to the jungle environment where they were born. Gradually the conservation agencies are returning the youngsters to their natural habitat. By 1967 Joan, the oldest adult female orang in the group, mated with a wild male and produced a baby, giving rise to optimism that the program could work. In the course of the program the baby orangutans exhibited many astonishingly human traits and attitudes—perhaps in imitation of the scientists and foresters who acted as their surrogate parents, as these pictures show.

When the orangutans involved in the Malaysian conservation program were reintroduced to their natural habitat, at first they seemed very unsure of themselves. These normally tree-dwelling animals spent much of their time on the ground (above, top) or nervously huddled together like the two youngsters above. After a while, however, the orangs become more secure in their surroundings and, like the ape at right, feel relaxed enough to sit back and leisurely pick termites off a twig.

In the wild, orangutans have been known to spend up to 60 percent of each day sleeping and napping. They often use their old nests for daytime snoozing and build fresh ones each night. Fatigue seems to have suddenly overcome the orang at left, however, for it simply curled up on a rock to catch a few winks.

To human observers it may seem that orangutans have a knack for doing some of the simplest things the hard way. The ape (below, left) executes a fancy over-the-head-behind-the-neck-with-stick maneuver to relieve an itch, while the hefty specimen directly below has to obscure its own vision in order to get to that tasty morsel on its pinky.

The Pampered Infancy of Orangs

Female orangutans have maternal instincts that are comparable to the mother love of humans. They dote on their babies. In the indiscriminate hunting of orangs that has made them the most endangered of the apes, most of the deaths have been of mothers defending their young. Orang babies are coddled and curried and indulged until they are weaned, which may be as long as four years. The mothers coach the babies in everything they must learn—stuffing premasticated food into their mouths to teach them to eat, pushing them out on branches to encourage climbing, leaving them on the ground to make them walk. By the time they are four, orangs can select their own food, like the fastidious youngster at left, daintily eating an orchid.

Getting the Hang of It

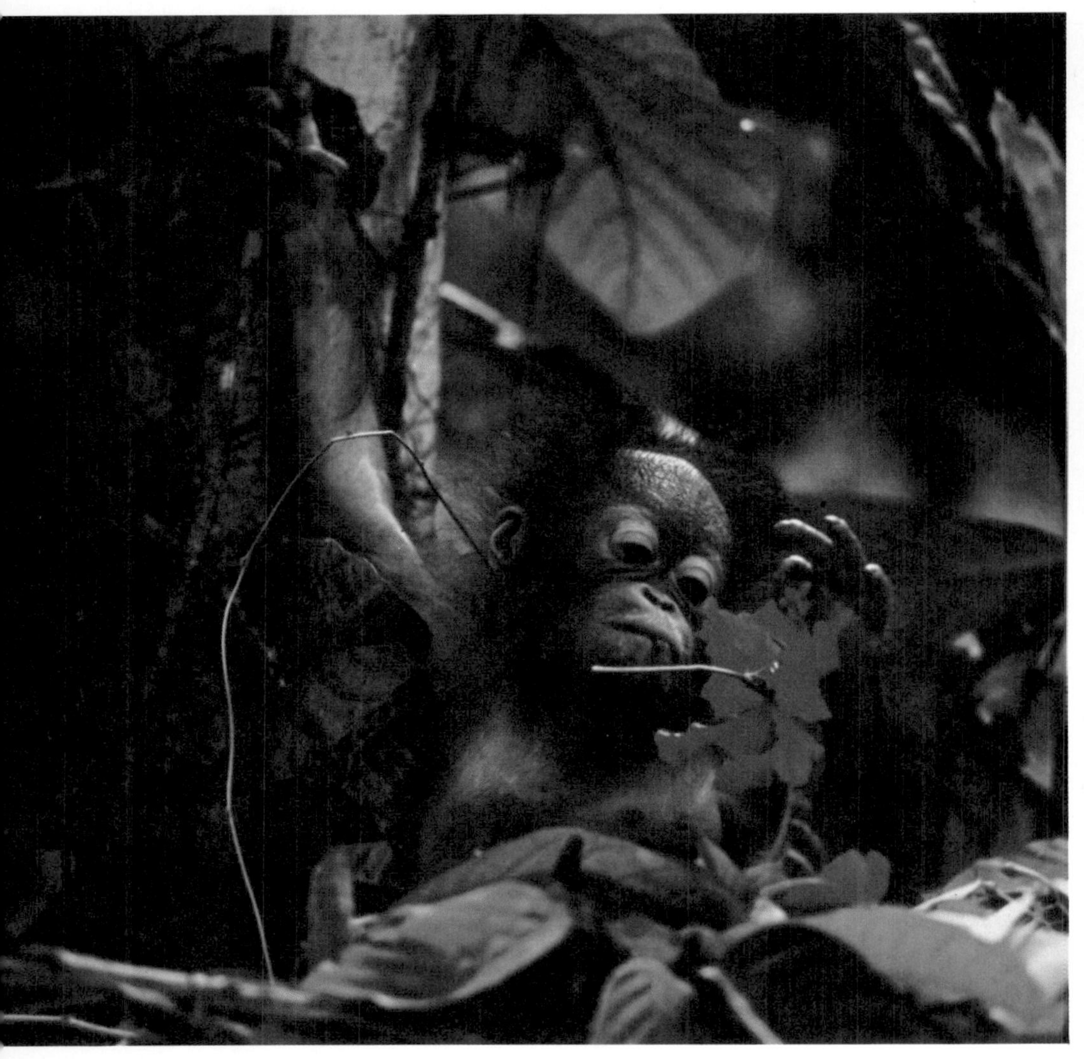

The two accomplished youngsters above are practicing the latest acrobatic turns they have learned, and the little fellow at left is already completely at home in the treetops. The hair-pulling match (top, left) helps teach these young apes about their own and each other's strength.

The Murders in the Rue Morgue

by Edgar Allan Poe

Edgar Allan Poe (1809–1849) was fascinated by the grotesque and unusual. In "The Murders in the Rue Morgue," he presented the reader with the first modern detective story, a baffling murder mystery offering a rational, although totally unpredictable, solution. The villainous orangutan of the tale is itself a victim of circumstance, driven to savagery and terror by its captivity.

Two women are found murdered and mutilated in an apartment in Paris. Witnesses, fairly consistent in their testimonies, all refer to a strange and shrill voice, speaking in a completely unrecognizable language, emanating from the room in which the murders took place. The bizarre and horrible facts provide detective-hero Auguste Dupin with the perfect material to test his exceptional analytic powers. After pondering the facts before him, Dupin inserts an advertisement in the newspaper stating that he has found an escaped orangutan and requests that the owner claim it. A sailor, Le Bon, shows up and tells the story that Dupin relates here.

What he stated was, in substance, this. He had lately made a voyage to the Indian Archipelago. A party, of which he formed one, landed at Borneo, and passed into the interior on an excursion of pleasure. Himself and a companion had captured the Ourang-Outang. This companion dying, the animal fell into his own exclusive possession. After great trouble, occasioned by the intractable ferocity of his captive during the home voyage, he at length succeeded in lodging it safely at his own residence in Paris, where, not to attract toward himself the unpleasant curiosity of his neighbors, he kept it carefully secluded, until such time as it should recover from a wound in the foot, received from a splinter on board ship. His ultimate design was to sell it.

Returning home from some sailors' frolic on the night, or rather in the morning, of the murder, he found the beast occupying his own bedroom, into which it had broken from a closet adjoining, where it had been, as was thought, securely confined. Razor in hand, and fully lathered, it was sitting before a looking-glass, attempting the operation of shaving, in which it had no doubt previously watched its master through the keyhole of the closet. Terrified at the sight of so dangerous a weapon in the possession of an animal so ferocious, and so well able to use it, the man, for some moments, was at a loss what to do. He had been accustomed, however, to quiet the creature, even in its fiercest moods, by the use of a whip, and to this he now resorted. Upon sight of it, the Ourang-Outang sprang at once through the door of the chamber, down the stairs, and thence, through a window, unfortunately open, into the street.

The Frenchman followed in despair; the ape, razor still in hand, occasionally stopping to look back and gesticulate at his pursuer, until the latter had nearly come up with it. It then again made off. In this manner the chase continued for a long time. The streets were profoundly quiet, as it was nearly three o'clock in the morning. In passing down an alley in the rear of the Rue Morgue, the fugitive's attention was arrested by a light gleaming from the open window of Madame L'Espanaye's chamber, in the fourth story of her house. Rushing to the building, it perceived the lightning-rod, clambered up with inconceivable agility, grasped the shutter, which was thrown fully back against the wall, and, by its means, swung directly upon the headboard of the bed. The whole feat did not occupy a minute. The shutter was kicked open again by the Ourang-Outang as it entered the room.

The sailor, in the meantime, was both rejoiced and perplexed. He had strong hopes of now recapturing the brute, as it could scarcely escape from the trap into which it had ventured, except by the rod, where it might be intercepted as it came down. On the other hand, there was much cause for anxiety as to what it might do in the house. This latter reflection urged the man still to follow the fugitive. A lightning-rod is ascended without difficulty, especially by a sailor; but, when he had arrived as high as the window, which lay far to his left, his career was stopped; the most that he could accomplish was to reach over so as to obtain a glimpse of the interior of the room. At this glimpse he nearly fell from his hold through excess of horror. Now it was that those hideous shrieks arose upon the night, which had startled from slumber the inmates of the Rue Morgue. Madame L'Espanaye and her daughter, habited in their night clothes, had apparently been occupied in arranging some papers. . . . The victims must have been sitting with their backs toward the window; and, from the time elapsing between the ingress of the beast and the screams, it seems probable that it was not immediately perceived. The flapping-to of the shutter would naturally have been attributed to the wind.

As the sailor looked in, the gigantic animal had seized Madame L'Espanaye by the hair (which was loose, as she had been combing it), and was flourishing the razor about her face, in imitation of the motions of a barber. The daughter lay prostrate and motionless; she had swooned. The screams and struggles of the old lady (during which the hair was torn from her head) had the effect of changing the probably pacific purposes of the Ourang-Outang into those of wrath. With one determined sweep of its muscular arm it nearly severed her head from her body. The sight of blood inflamed its anger into phrensy. Gnashing its teeth, and flashing fire from its eyes, it flew upon the body of the girl, and imbedded its fearful talons in her throat, retaining its grasp until she expired. Its wandering and wild glances fell at this moment upon the head of the bed, over which the face of its master, rigid with horror, was just discernible. The fury of the beast, who no doubt bore still in mind the dreaded whip, was instantly converted into fear. Conscious of having deserved punishment, it seemed desirous of concealing its bloody deeds, and skipped about the chamber in an agony of nervous agitation; throwing down and breaking the furniture as it moved, and dragging the bed from the bedstead. In conclusion, it seized first the corpse of the daughter, and thrust it up the chimney, as it was found; then that of the old lady, which it immediately hurled through the window headlong.

As the ape approached the casement with its mutilated burden, the sailor shrank aghast to the rod, and, rather gliding than clambering down it, hurried at once home—dreading the consequences of the butchery, and gladly abandoning, in his terror, all solicitude about the fate of the Ourang-Outang. The words heard by the party upon the staircase were the Frenchman's exclamations of horror and affright, commingled with the fiendish jabberings of the brute.

I have scarcely any thing to add. The Ourang-Outang must have escaped from the chamber, by the rod, just before the breaking of the door. It must have closed the window as it passed through it. It was subsequently caught by the owner himself, who obtained for it a very large sum at the *Jardin des Plantes*.

The Airborne Orangutan

On the ground, orangutans are awkward, slow-moving creatures, but in the trees they are completely at home. The youngsters swing nonchalantly along the jungle canopy with a grace and rhythm that have been likened to a stride, or simply hang from a branch (left), plucking the fruit and foliage they eat or just relaxing. The ability to swing through the trees and vines is not innate, though, and young orangs must be painstakingly coached by their mothers before they can take to the trees with confidence.

Until they reach sexual maturity at the age of 10, orangutans are lively, playful creatures, but with adulthood they lose most of their youthful exuberance and gradually slow down, performing their aerial feats only occasionally and spending most of their time sleeping, napping or sitting.

Gibbons

Because their size and their method of moving about are so distinctly different from the other apes, the gibbon and the closely related siamang are classified by zoologists as a distinct subfamily. They are called *Hylobates*, a Greek word meaning "tree-walkers."

Unlike the heavyweight gorilla, chimpanzee and orangutan, the gibbon weighs only 10 to 20 pounds and the siamang only 20 to 30, although their dense, fluffy fur makes them look much bigger. Unlike the other apes, too, gibbons have a distinct anatomical feature—tough, horny pads on their buttocks fused to the hip bones. These sitting pads, known as ischial callosities, are indispensable, for the gibbon builds no sleeping nest, and the pads guarantee it a comfortable night seated on the narrow branches where it can feel safe from leopards and other predators.

The gibbon's lightweight physique suits it well for its aerial environment, high in the forest canopy. Gibbons spend much of their lives 100 to 150 feet above the forest floor, where fruit, birds and eggs are most abundant. From branch to branch they swing along lightly and effortlessly, hand over hand. Their long arms give them the locomotor equivalent of a long stride, but gibbons improve on this by letting go between handholds, literally flying through the air with apparent reckless disregard for the law of gravity. This hand-over-hand treetop locomotion is called brachiation and is used only occasionally by chimpanzees, orangutans and young gorillas.

Gibbons are found throughout the tropical rain forests of Southeast Asia and the Indonesian islands of Sumatra, Java and Borneo. They live in "family" groups, which include an adult male and female and their offspring. Each group occupies a living space or territory in the forest, but because of the unpredictable nature of the fruit harvest in different years, the group establishes as large a territory as it can and tries to prevent other gibbon groups from encroaching on it. There are continual conflicts over boundaries, especially when a fruit tree ripens in the common ground between two territories. But gibbons rarely get hurt, because these conflicts are ritualized, and noisy mock battles take the place of real fighting.

One of the most characteristic early-morning sounds in gibbon country is the swooping high-pitched song of the female. The clear hooting notes begin slowly, then swell and accelerate until they coalesce in a bubbling coloratura trill of tremendous carrying power. From the distance comes the answering call of a neighboring female, and the two families quickly converge on their common boundary. Now the males rush at one another, chasing and dodging among the branches, or, hanging by one hand, confront one another, hooting menacingly. Throughout the mock battle the females encourage the males by their loud calls. After about an hour the groups move back into their own trees, and the episode is over. Once again the frontier between them has been defined.

The newborn gibbon is naked at birth except for a little furry cap. Its mother bends her knees up to her chest to support it, making a snug nest. Its skinny pink arms may look wretchedly feeble, but its hands and feet have a tenacious grip on its mother's fur from the first day of life. The mother can swing through the trees freely, evidently without concern for the infant's safety.

By the age of two the baby gibbon is independent but still very much part of the family. Then a second baby is born and then another, at about two-year intervals. By the time the fourth baby comes on the scene, the eldest, by now about six years old, is approaching maturity—and arousing feelings of antagonism in the breast of one of its parents. Adult gibbons feel hostile toward other adults of the same sex whether they live in a neighboring group or whether they are part of the same family. The result is that the adolescent is gradually eased out of the group and leads a solitary life until it meets and pairs with another adolescent.

The siamang, from Malaya and Sumatra, shares many behavioral characteristics with the other gibbons—for instance, the "family" type of social group. They even share the same territory comfortably, as the gibbon is mainly a fruit-eater while the siamang is predominantly a leaf-eater. The siamang, equipped with a special throat sac, is able to emit a call that is even louder than that of the other gibbons. Gibbons and siamangs must surely be among the noisiest mammals in the world.

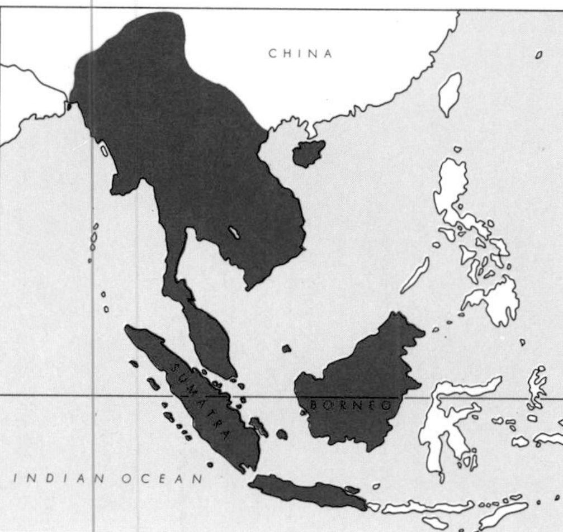

Gibbons are found throughout the tropical forests of Southeast Asia and Indonesia.

"...With the Greatest of Ease"

Gibbons are the most skilled brachiators among the apes. In the tropical forests where gibbons live, the trees grow to immense heights. The branches thrust upward toward the sky, with the shoots of neighboring trees crisscrossing each other, forming numerous horizontal strata. Climbing vines hug the tree trunks and coil around the branches, creating the chainlike bonds between the trees that are the equipment needed for the apes to perform their arboreal acrobatics. Hanging by their exceedingly strong, slender arms and using their long, tapering fingers like hooks, the gibbons swing through the trees, sometimes leaping distances of 30 feet and more between the branches.

In a breathtaking display, the two young gibbons on these pages vividly illustrate why these apes are unsurpassed in their ability as treetop locomotors. Even the most agile monkeys would hesitate to traverse a vine on anything less than all fours. But the gibbon makes the trip using any and all combinations of its limbs. Sometimes an ape (opposite, above) will even balance dramatically on the liana vine without using its hands at all, seemingly unperturbed by the dizzying distance between itself and the forest floor below. Whether in the trees or on the ground, the gibbon raises its arms gracefully at its sides when it is walking erect. This helps maintain equilibrium and, on the ground, keeps those disproportionately long limbs from dragging.

The Gibbon in China by R. H. van Gulik

The gibbon frequently appears in the literature of China, where it was often raised as a pet. One of the most famous tales of a man and a gibbon was first recorded at the end of the T'ang Dynasty (10th century). Although the original manuscript, Experiences of Wang Jeñ-yü, *has not been preserved, the adventures of General Wang Jên-yü and his pet gibbon were retold in* The Gibbon in China *by R. H. van Gulik, from which the following excerpt was taken. The setting is the northern frontier of Szechuan, a mountainous region where there is a large gibbon population. The description of Yeh-pin's mischievousness, his cleverness and loyalty are an accurate portrayal of the behavior of domesticated gibbons.*

When Wang Jên-yü was serving in Han-chung, he had established his household in the office compound. A hunter from the Pa mountains presented a young gibbon to him. Wang took pity on it because it was so tiny and so clever, and had his men raise it, bestowing upon it the name of Yeh-pin. If he called the gibbon, he would respond at once. After some years the gibbon had grown strong and tall, and had to be kept on a chain most of the time, for he would bite everyone who came near him, thereby causing his master much worry. Everytime Wang scolded the gibbon, he would be good again and behave well for a time. But the gibbon did not fear anybody else, even when they came armed with a whip or stick.

Wang's residence was surrounded by walls, and elms and locust trees grew there. In the temple dedicated to Emperor Kao-tsu of the Han Dynasty stood tall pinetrees and old cedars where countless birds had built their nests. On the day of the full moon of the second month, Yeh-pin got loose. He jumped into the forest and crashed about in the trees. Then the gibbon entered the temple-compound and threw the young birds and eggs onto the ground. In front of the Prefect's office stood a rack carrying small bells on strings [i.e., a kind of burglar alarm—V.G.]. The bells began to ring when a flight of birds alighted on the strings. When the officer in charge went to investigate where those birds came from, he saw Yeh-pin in a tree, and at once ordered his men to pelt him with stones and shoot arrows at him. No one, however, could hit the gibbon. Only towards night fall when the gibbon's stomach was empty, did hunger drive him down so that he could be put on his chain again. [After these mishaps] Wang ordered his men to set Yeh-pin free in the Pa mountains, a hundred miles away. Just when the men had come back, and before Wang had even finished questioning them how things had gone, Yeh-pin was already back in Wang's kitchen, trying to snatch some food.

Wang again put him on a chain, but one day the gibbon again got loose. He went to the commander's kitchen, threw the kitchenware about, breaking and dirtying bowls and dishes. Then he climbed onto the roof, tore up tiles and bricks and threw them down. The commander flew into a violent rage and ordered his men to shoot the gibbon. But Yeh-pin just remained sitting on the ridge of the roof and continued to tear off the tiles. Although there was a rain of arrows, he remained completely unconcerned: he caught some with his hands and pushed other arrows away with his feet, dodging them so cleverly that the arrows flew past him right and left without harming one hair of his fur. Then an elderly officer called Ma Yüan-chang said that in the market there was an experienced monkey-keeper, and that man was called. The man pointed out the gibbon on the roof to his macaques, telling the monkeys to catch him and bring him down. Thereupon a few big macaques jumped onto the roof of the office, and went for Yeh-pin. The gibbon sprang down over the wall into the street and there he was caught. When brought before the commander, Yeh-pin was drenched with sweat and seemed most humble and repentant. The commander did not scold him too severely, and all who witnessed the proceedings laughed heartily. Wang Jên-yü tied a red ribbon around Yeh-pin's neck and wrote on it the following parting poem:

I set you free, enjoining you to return to your former forest,
You will easily find your way back to your cherished haunts.
When the moon shines on the Wu Pass, you will enjoy its quiet,
And the trails across the Pa mountains you will not find too steep.
Tarrying there you will not be plagued by dreams about green mountains,
Climbing high you will enjoy yourself deep in the azure clouds.
When in the three months of autumn fruit ripens and the pines are strong,
Carelessly clinging to a high branch, your morning song'll be heard far and wide.

Thereupon Wang ordered his men to take the gibbon to the Ku-yun-liang-chueh mountain, telling them to keep him tied up in a farmhouse there (so as to make him forget his old home). After ten days they undid the gibbon's chain and set him free. This time Yeh-pin indeed did not come back.

Later Wang was relieved of his duties in Han-chung and crossed over into Szechuan Province. When he and his retinue halted in front of a temple on the Po-chung mountain, on the bank of the Han river, a troop of gibbons let themselves down, holding each other's hands and feet, to drink from the clear stream. Then a large gibbon left the troop and came forward. Hanging from the branch of an old tree by the roadside, he watched the men below; he still had the same red ribbon round his neck. A member of Wang's retinue pointed at the gibbon and said "That's Yeh-pin!" Wang called him, and the gibbon called out in reply. When they mounted their horses the gibbon seemed sad, and when they pulled the reins the animal uttered a mournful wail, then disappeared. But when Wang's cortege was ascending the mountain road that wound upwards among valleys and brooks, they heard all the time a sad wailing from the trees by the roadside, and they supposed it was the gibbon who was following them, in great grief. Then Wang composed a sequel to his former poem, saying:

In front of the Po-chung Temple, on the bank of the Han river,
These gibbons, strung together, descended from a high cliff.
When they came near for a closer look at the travellers,
I recognized Yeh-pin, who was looking exactly as before.
Resting in the moonlight, he won't be plagued by dreams of chain and leash,
Feeding on pineseeds, he has shed the cereal-eater's body.
He called a few times, the heart-rending sound rising to the clouds:
He had indeed recognized his master of bygone years!

In a Family by Themselves

There are seven species of gibbons, including the siamang. Gibbons are classified as a distinct family group among the primates. They are totally tree-dwelling, and their ability to maneuver among the vines and branches enables them to elude predators and to tap the abundant supply of figs, grapes, mangoes and other fruits that make up most of their diet. The individual members of each family group are very close to and protective of one another, and, seeking safety in numbers, they huddle together in a tree in groups of two or three while they sleep.

Gibbons, like the one above (left), don't build nests but sleep sitting upright, resting comfortably on their tough sitting pads in the fork of a tree. The fact that the ape at right (above) is a captive siamang doesn't inhibit it from inflating its throat sac and proclaiming its territory with a booming "wow." The hoolock gibbon (right) is distinguished by a white band of fur across its forehead. Its body hair is so long and thick that it covers the ape's ischial callosities.

Langurs and Colobus Monkeys

Old World Monkeys are divided into two subfamilies, classified by their diets: the omnivorous monkeys, which eat anything from fruit and insects to the flesh of small animals, and the leaf-eaters, which eat only tropical foliage. Leaf-eaters in turn come in two types, the langurs of Southeast Asia and the colobus monkeys of Africa. Because of their fastidious appetites and the difficulty of finding suitable substitutes for their natural food, the leaf-eaters are rarely seen in zoos and are lesser known than their omnivore cousins. They are, nevertheless, a large, thriving and wide-ranging group of primates. Their digestive tract has much in common with that of a ruminating animal such as a cow. Whereas a cow's stomach is virtually a chemical factory, with four compartments designed to process a bulk intake of grass, the langur's partially divided stomach can deal with large quantities of mature leaves. In tropical forests leaf-eating monkeys find a constant supply of food all around them, so there is no need to raid crops, and, significantly, they lack the capacious cheek pouches, the shopping bags of macaques, mangabeys, baboons and guenons.

The name *langur* comes from the Hindustani *lungoor* (meaning "long-tailed") and, strictly speaking, should be used only for the Hanuman, or entellus, langur, which is found throughout India from the snowy regions of Kashmir and Nepal to Ceylon. Hanuman langurs are held sacred and are never molested by Hindus because they are thought to be descended from the monkey-god Hanuman, who aided the God Rama and the Goddess Sita in a series of legendary exploits recounted in Indian mythology.

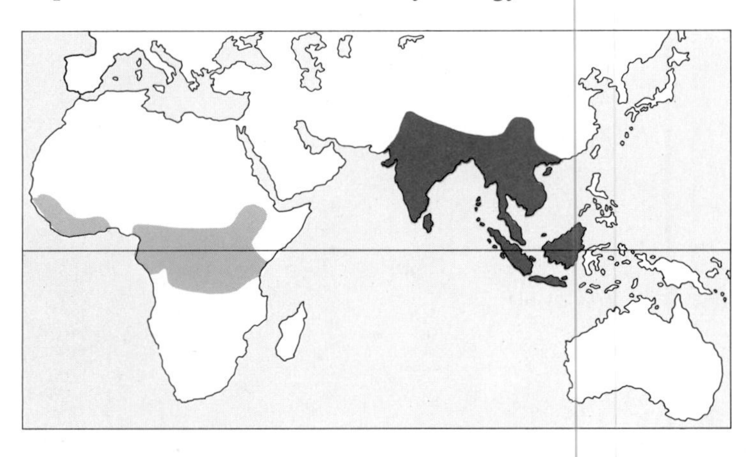

Langurs (red) range from India to Indonesia. Colobus monkeys (blue) live in central and western Africa.

Hanuman langurs live in groups of about 20. Although there may be four adult males and eight adult females in the group, it is the infants who seem to exert the main cohesive force. At birth the infant, pink-skinned and sparsely covered with black fur, is inspected thoroughly by its mother, who licks it all over, grooming and stroking it gently. It immediately becomes the center of attention of all the females in the group. As soon as it is dry, its mother allows one of the females to take it from her for further inspection, licking and nuzzling. After a while the baby sets up a wail of protest and is passed on to yet another eager female, who goes through the same routine. The mother will retrieve it if its squeals become too anguished, but it may well be handled by seven or eight strange females during the course of its first day of life.

The care of adult females for infants other than their own is known as "aunt" behavior and is taken to great lengths by the langurs. The contrast between the black fur of the newborn infant and the creamy coat of the mother probably acts as a signal that elicits care and protection for the infant at the time when it needs it most. The handling of infants is also of adaptive value since it means that a langur female, by the time she comes to have an infant of her own, has had a good deal of experience. When the infants are about three months old langurs make baby-sitting arrangements whereby two or three mothers go off to feed, leaving their infants in the care of an "aunt."

Colobus monkeys are unique in the Old World since they lack one of the most significant primate characteristics—a thumb. The Greek word *kolobus*—meaning mutilated or stunted—implies an animal that is deprived of a vital manipulative skill. Yet the colobus is in complete control as it bounds through the topmost branches of the forest, 100 feet above the ground. Its hands may be mere hooks, with the four fingers hinging over to meet the palm like a spring clip, but they are ideal for cavorting in the tops of trees and equally effective when it comes to feeding. The colobus has only to bend a branch and pick off the leaves with its lips.

Finally, in the Far East are found what are known familiarly as the "funny-nosed" leaf-eaters, who, by some quirk of natural selection, have acquired either too little nose or too much. From western China comes a chocolate-and-gold monkey with the minutest of snub noses set in a blue face. The oriental-eyed douc from Vietnam (opposite) has even less nose—just two oblique slits.

The Holy Monkeys

Hanuman langurs are revered in India and have become so accustomed to coexisting with humans, secure in their sacred status, that they invade cities and plantations with impunity, begging for food and stealing from orchards unmolested. Langurs display a spectacular variety of markings, like the dusky langur at left, which was bright orange at birth and acquired its darker coloration and distinctive white "spectacles" at the age of six weeks. What all langurs have in common are their slender, lithe physiques, naked faces and special digestive systems that enable them to subsist on a diet of leaves. Apart from its cousins, the colobus monkeys of Africa, the langur family occurs only in southern Asia, in a wide variety of habitats, ranging from 12,000 feet on the slopes of the Himalayas to the tropical rain forests of Malaysia.

Because of their leafy diets and complicated digestive systems, langurs must sit quietly in the crowns of trees for hours at a time, digesting their food, like the Hanuman langurs above. They are not sedentary all the time, though, and can scamper around their arboreal habitat or across the ground (opposite) with agility and speed, using their extra-long tails as balancing stabilizers.

At birth, most langurs have strikingly different colors and markings from their parents and change to conform with the family coloration within a few weeks. The baby Hanuman langur at right, clinging to its mother's arm, was dark brown when it was born and will quickly change its colors to conform with the adult's platinum-blond coat. According to Hindu legend, the sacred Hanuman monkeys got their distinctive black faces and paws when their ancestor rescued the queen-goddess Sita from an evil monster, which thereupon tried to burn the monkey at the stake. But the clever Hanuman escaped, with only its scorched face and paws as a memento.

The Flying Colobus Monkeys

Several species of the long-haired colobus monkeys were brought to the verge of extermination in the past by hunters seeking their luxuriant silky fur, especially the flowing flank fringes and plumed "horsetails" of some subspecies. To satisfy the fashion trade, 170,000 black-and-white colobus monkey pelts were exported from Africa in 1892. Fortunately for the colobus monkeys, the vogue for "monkey fur" passed after World War I, and recent generations of the family have come back from near-extinction. But one species, the wistful-looking, tousel-haired red, or Zanzibar, colobus (left), has been the victim of another kind of enemy and is now nearly extinct. Intensive cultivation and settlement of its habitat on the island of Zanzibar has reduced the number of red colobus monkeys to a few hundred.

Living in the tropical forests and mountains of Equatorial Africa, colobus monkeys are incredibly able and daring aerialists, as shown in the filmstrip at right. They will change direction in mid-air (above) or deliberately take dives of 20 or 30 feet to the branches of middle-level trees.

The Proboscis Monkey–Nosiest of the Simian Family

The adult male proboscis monkey is the Cyrano of the monkey family. There is nothing quite like his huge comical nose among the other primates. It sometimes reaches a length of three or more inches and keeps on growing even after the male reaches adulthood. Scientists believe that the masculine nose is the result of selective adaptation: The females prefer to mate with big-nosed males. Females and youngsters have smaller, upturned noses that are about the size of man's and are similar to the noses of their cousins, the snub-nosed monkeys. In practice, the male proboscis monkey uses his nose as a sound amplifier when he makes his ritualistic territorial cries.

Proboscis monkeys are found only on the island of Borneo and are becoming increasingly rare. Like other langurs, they are superb climbers and fearless sky divers and will unhesitatingly take off with arms outstretched, like a free-flying glider (above), to the branches of a distant tree on a lower platform of the forest.

Guenons

The most numerous forest monkeys of Africa, the guenons, are renowned for the variety of their colors, spots and stripes. Slightly smaller than macaques—about one and a half to two feet—they run swiftly along the aerial pathways of their forest home, using their long (up to three and a half feet) tails as balancing aids. The precision of their markings satisfies the need to recognize their own and other species easily, even when only a small portion of a body can be seen through a screen of foliage. The distinctive markings of guenons are concentrated mainly on the face, rump and legs. White bibs, white nose blobs, white whiskers, mustaches and beards and white tiaras are complemented by white rump patches and thigh stripes. A chestnut tiara, blue facial skin, white beard and white thigh stripe give distinctive dignity to the De Brazza's monkey from the Congo Basin (opposite); the snow-white collar and diadem, the black face, dark-gray coat and red thighs of the Diana monkey from West Africa have great elegance. As all monkeys of the Old World have excellent color vision, instant recognition of species must be very important. An adult male guenon, for example, looking at his own reflection in a mirror hanging in a tree, will threaten his own image, bobbing his head up and down in exactly the same menacing way as if it were actually confronting a stranger.

Most forest guenons live in small groups, typically one adult male, two or three adult females and as many as eight offspring. The male will not tolerate another adult male, hence the head-bobbing threat signal, but he is not as bossy and aggressive as the male macaque. A relaxed atmosphere prevails as the group moves, feeds on everything from fruits to lizards and grooms in the safety of the trees, though the patriarch remains on the alert for snakes, birds of prey or other predators. Young guenons indulge in chasing and wrestling as young primates do the world over. They even invent games of their own, such as waiting for a flock of solemn hornbills to get nicely settled in the branches of a tree and then shaking the branches and making the birds fly off in panic.

Another guenon, the tiny talapoin monkey, has a unique life-style for simians. In strips of flooded forest along the rivers that seam the equatorial jungle, it feeds on the roots of the manioc or cassava plant that villagers have gathered and left to soak in riverside pools. The talapoins, attracted by ample food, congregate near villages in groups of more than 100, leading a semiparasitic existence, not unlike the rats and mice of our own barns and granaries. When freshly gathered, the manioc root is poisonous and very bitter; it has to be soaked in water for several days to leach out the poison by fermentation before it is safe to eat. In making manioc their staple diet, the talapoins have raised an interesting scientific question. How, when they may choose among thousands of manioc roots at various stages of fermentation, can talapoins tell which are nonpoisonous?

Some guenons have established themselves on forest fringes where clumps of trees and bushy thickets are interspersed with wide expanses of grassland. A strip of evergreen forest growing along a watercourse in otherwise open country is a favorite habitat of the savanna monkey, which is found from Senegal in West Africa to Ethiopia in the east and southward to the Cape of Good Hope. In these three localities it is called the green monkey, the grivet and the vervet, respectively.

One guenon—the patas monkey—has left the forest altogether and lives full time in the savanna. It is probably the fastest primate on earth, built like a greyhound, with long legs that give it a tremendous loping stride. With a russet-red coat, gray chin whiskers and white military mustache, it looks like a grumpy, retired British colonel. Patas monkeys live in groups of about 15 with only a single, dominant, fully adult male. The patas male plays a role that is well suited to the open country where the group is exposed to attack by leopards and hyenas. He acts as a watchdog, standing on two feet to peer over the tall grass, sometimes using his tail as a tripod, or climbing into an isolated tree to spy out the land. If he sees a predator, he utters no noisy alarm bark, just a soft chirruping call that alerts the group. Then they crouch silently in the grass, remaining concealed while the male performs a conspicuous diversionary display. Bouncing noisily about in the branches of his tree, he rushes off in the opposite direction, giving time for the females and young of the group to make their escape.

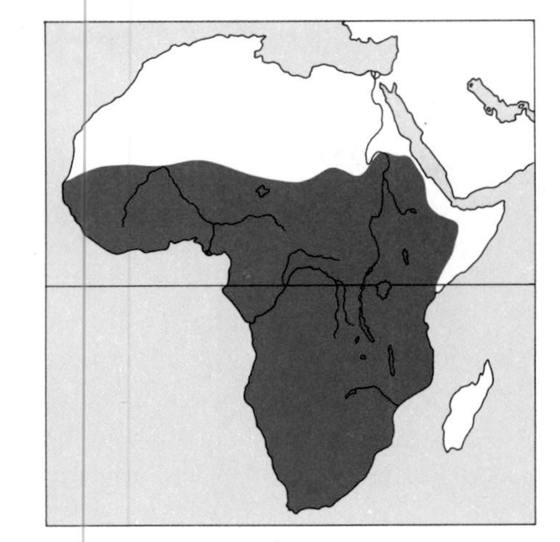

The colorful guenons inhabit the forests and savannas of Africa south of the Sahara Desert.

A Colorful Family

Guenons, the common, colorful monkeys of Africa, include about 15 different species, most of which inhabit the forests south of the Sahara. The spot-nosed guenon (above, left) lives mainly in the middle layers of the trees, occasionally attracted to the topmost canopy by fruits and flowers. Unlike the other guenons, whose calls are usually noisy, these monkeys communicate with a few soft birdlike chirps and twitters. The patas monkey (above, right) avoids forests completely, preferring the almost treeless, grassy plains of the African steppe. Even in the presence of danger, the swift patas monkey can almost always make a successful getaway on the ground. The handsome Diana monkey (opposite) lives in tropical forests from Sierra Leone to Ghana. In addition to their stunning white beards, these active and curious animals have dramatic red slashes of color on their thighs.

Although the grass monkeys are primarily savanna-dwellers, they do spend a portion of their time in the trees. For the vervets (opposite) the trees provide a good hiding place from predators. They blend so well with their surroundings—the tail of the topmost monkey so still and branchlike—that they could easily go unnoticed. The vervet youngsters above go out on a limb waiting for some motherly grooming. The lone vervet at right, however, simply relaxes on a branch, a comfortable enough perch thanks to its small but cushioning sitting pads.

The Grass Monkeys

The grivets, vervets and green monkeys are guenons that are known collectively as grass monkeys. No matter what their name, these animals all tend to avoid waterless, arid areas. They choose instead the vicinity of rivers, especially the banks, as well as the boundary regions of the rain forests where fresh water is always available. They live in groups that, if living conditions are favorable, can have as many as 50 members. But, like their forest-dwelling relatives, they split up into smaller troops to forage for the plants, nuts, fruits, insects and even small animals on which they feed.

Baboons

Tough, aggressive, adaptable—that's the baboon, and it must be all of that to survive. It inhabits almost every kind of environment in Africa and the southwest corner of Arabia, from evergreen forest to grassland, even parts of the Sahara Desert. Roaming the open savannas from the rocky outcrops of Ethiopia to the treeless *kopjes* (small hills) of the south, it often experiences hunger and thirst and must be constantly on the alert against predators.

Baboons are the largest monkeys of the Old World. The males of some species measure over three feet from muzzle to rump and weigh up to 90 pounds, the females only half as much. They are omnivorous, sometimes traveling 12 miles a day foraging for food, their dull speckled coats blending into a background of dry grasses and acacia scrub. When fruit is not available they settle for grasses and seeds. They dig for roots and tubers, overturn stones to find insects and catch lizards and occasionally even hares and newborn gazelles. Under the blazing African sun, water is of vital importance. A waterhole may be shared with several other baboon troops as well as impalas, zebras and wildebeests. In intense drought baboons find water by digging holes in the sand of dried-up riverbeds.

In open country there is the constant threat of attack by such predators as lions, leopards, cheetahs, hyenas, jackals and wild dogs. At night baboons seek the protection of trees as sleeping sites, or, in treeless regions, steep cliffs where they cannot be easily pursued. Baboons sleep sitting upright, and, like gibbons, they are equipped with ischial callosities, the leathery sitting pads that make it possible for them to perch comfortably.

Baboons live and travel in large, tight-knit groups, a defense system that is tremendously advantageous in open country. Several pairs of keen eyes scan the surrounding terrain on the alert for any surprise attack. An alarm bark warns the troop, which instinctively bunches together, with the females and young in the middle and the young males on the periphery. Then four or five adult males advance in a phalanx toward the source of the trouble. Their imposing capes of bristling shoulder hair and their flashing canine teeth act as a deterrent and discourage a frontal attack. Most predators prefer to pick off a straggler, such as an inexperienced juvenile or a sick animal that cannot keep up with the troop, rather than confront the main body.

A soothing activity that unites all members of the baboon group in friendly relaxation is grooming. The adult females play the major role, grooming infants, particularly their own offspring, as well as juveniles, adult males and one another. The baboon's hand is well adapted for this delicate occupation, parting the hair and picking off burs, scabs or other foreign matter.

Troops of savanna-living baboons average 40 in number, including six adult males, 12 adult females and about 20 assorted offspring. In this type of group each adult male may mate with each adult female. Hamadryas baboons, living in the arid environment of East Africa and parts of southern Arabia, have a different arrangement. Each adult male has exclusive rights to a "harem" of one to four females, not just during the estrus period but throughout the year. While each unit forages independently during the day, they reassemble at night, since in their inhospitable habitat there are few suitable sleeping sites. Each evening many one-male units come together into a large herd of about 100 strong. The male guards his females jealously, threatening them if they stray too far away. As dusk falls and the herd climbs to the safety of the precipitous cliffs, each "harem" unit huddles together to sleep. Another group of baboons with stumpy tails lives in the equatorial forests of the west coast of Africa. These are the mandrills and drills, renowned for their spectacularly colored bare bottoms, which are sky-blue, violet and rose-pink. Little is known of their way of life deep in the recesses of the forest, but large herds of over 100 have been encountered.

From the time explorers first visited Africa they have returned with tales about how baboons attack people by deliberately rolling stones down on them from cliffs or hillsides. Scientists have consistently pooh-poohed such yarns. Recently, however, three Stanford University anthropologists were compelled to revise their ideas when large stones whizzed uncomfortably close to their heads. The culprits, caught in the act, were a group of baboons.

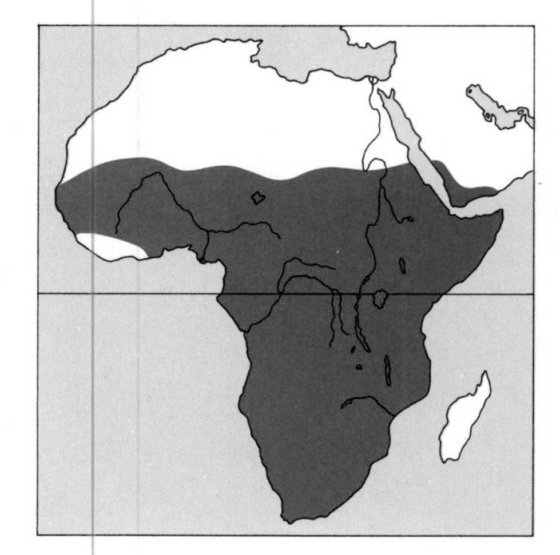

Baboons roam the savannas, forests and rocky plateaus of Africa and southernmost Arabia.

The Organization Baboon

Like his human counterpart, the organization man, a baboon must learn to "play the game" if he expects to succeed in the troop. And the game has one all-pervasive rule: Submit to those who are stronger, dominate those who are weaker. A rigid adherence to this role-play is vital to the survival of the troop, for it permits them to range for food over a far wider territory than an individual baboon would dare cover and to make the most effective use of their combined strengths in repulsing the attacks of predators.

A male baboon encounters this dominance-submission theme from the moment he leaves his mother to join his first play group; fighting and jockeying for a more dominant position is constant. Inevitably, a young male's ability to assert himself already has been affected by the behavior he observed in his mother. If she was self-assured and dominant, the chances are better that he will be the same.

In early adulthood he continues his battle toward the top by combining sheer muscle power, the ready use of sharp canine teeth and a complex system of threatening gestures and facial expressions intended to intimidate an opponent without a fight. Eventually the aspirant may win his way onto the management team, a small group of dominant males who cooperate in ruling and protecting the troop against both internal disorder and outside threats.

A steely stare is the opening gambit in this confrontation between two gelada baboon males (left). Most squabbles are ended by such symbolic threats rather than by fighting. On the road during food hunts (right) dominant males travel in the troop's center with the females and infants, while younger males range ahead and to the sides, alert for danger. At a site of food (below)—usually grass, roots or fruit—the scene is an active swirl of digging, eating, grooming, playing and roughhousing.

ANIMAL

Born in bleak, cold Edinburgh, Ivan T. Sanderson longed for a hot and sunny clime, and at the age of seventeen he journeyed to Malaya to collect animals. He came to realize, however, that simply collecting specimens was an outdated approach to zoology, and he developed an interest in the behavior of animals in their natural surroundings. He arranged an expedition to the southeastern corner of Nigeria, an area that had received scant attention from other naturalists. It was on that expedition that Sanderson met a group of baboons, an experience he affectionately and amusingly describes below.

Suddenly before me sat a most menacing figure, apparently wrapped in a grey shawl, and scrutinizing me with a pair of unpleasant-looking eyes from beneath a scowling brow. He (or she) and I both ceased our visceral mutterings promptly and uttered a surprised "uh" so precisely in unison that I got an overpowering desire to giggle. This was, however, as quickly wafted away also, when my zoological reasoning came to an abrupt halt. I had not suspected that drills (*Cynocephalus leucophæus*), though baboons of a sort, went about in large, belching parties.

Now I had met baboons before and although I took a great interest in their behaviour and would have liked to have returned to camp with a fine specimen such as now sat complacently before me, I remembered that discretion was always the better part of valour when in their presence. I therefore stood up to go, trying to be as unhurried as I imagined I would be at a vicarage tea party, though I have never attended one. This simple movement, however, was heralded by unmistakable complaints from all sides in the form of the most unpleasant grunts. The old lady (or gentleman) before me also rose, but on all fours so that his or her posterior came into view. It happened to be bright pink at this time of the year, and I thought absurdly of the homeric description of the dawn as "rosy-fingered."

This display had remarkable effects. The bushes parted on all sides and a surprising array of subhumanity presented itself, ranging from one obvious male of quite alarming proportions, to the merest toddlers with pale, flat faces quite unlike their dog-nosed, black-visaged elders and betters. Their movements were leisurely, as if they were taking their places for a boxing match; they chattered and grunted exactly like any crowd of pleasure-seeking human beings preparing for an entertaining display.

While all this taking of seats was going on, I was retreating gingerly backwards up the path, while trying to learn the rules of monkey ethics in the raw. The outsize gentleman seemed to have been appointed as doorman. He trotted into the path behind me and stood squarely upon three boulders, one for each back foot and one for his gnarled hands. This was all very unpleasant and I found myself waiting with some trepidation to see what was the next item on the agenda. As they continued to sit and grunt to each other, it appeared to be up to me.

I don't expect you have ever been surrounded by a troop of expectant baboons, but if you have, you will probably agree that it becomes extremely difficult to think up any parlour tricks. My mind was a blank, especially as each part of the circle to which my back was turned in succession seemed to think its chance had come to grab a ringside seat, and since one can't face all ways at once, the ring began to diminish rapidly. I remember thinking stoically and hopefully that drills are vegetable feeders and that I was not a vegetable although I doubtless looked like one. When the old gentleman yawned, and I had a glimpse of his three-inch fangs, I began to doubt the words of wisdom uttered by the worthy professor of my late and, at that moment, greatly lamented university.

I did remember that almost any animal, even a surprised tiger, will shy away if one stoops to pick up a stone and makes pretence of throwing it. This I instantly put to the test, but in my excitement I accidentally did pick up a

TREASURE *by Ivan T. Sanderson*

stone and hurled it at the big yawning male with a force of which I did not believe myself capable. We were all greatly surprised when it found its mark in a glancing blow. This, combined with my sudden action, made the spectators jump backwards with some emphasis so that I was given quite a lot of room to move about in. My target seemed quite angry, as might be expected, and as I stooped to gather more missles, he waltzed about and returned the compliment with some vigour, scraping the ground with his hind feet, gathering up a small boulder in the process, and projecting it straight at me with considerable accuracy.

This heralded a great commotion. Apparently the show had begun. I hurled more stones in all directions, and although the admiring onlookers retreated each time, those on the opposite side advanced, the gentleman who had yawned so indulgently most of all. He was now very angry indeed, projecting stones and big blobs of spittle at me alternately as he waltzed about, presenting first his revolting, dog-like visage and then his still more revolting and quite uncanine other end.

These tactics, combined with more stones and the fast-descending dusk, made me not only definitely frightened, but inexplicably angry too. Once I nearly put a charge of lead into him, but luckily checked myself, realizing that this trump-card would be even more useful later on when the difference of opinion became general. Matters appeared to be rapidly drifting in that direction.

During one of the periodic lulls between these diplomatic interchanges, now carried on in a more or less tense silence, one of the smallest and most youthful of my audience uttered a peevish squeal and bowled a small lump of earth at me, just as an underhanded lob-bowler in a juvenile cricket match would do. The action was so ludicrous that in my decidedly agitated frame of mind I burst into roars of laughter. Why it seemed so screamingly funny I don't know; perhaps it wasn't really so at all. But my action proved a most fortunate one.

The brat's mother made a dive at her now cowering and shivering prodigy, gathered it to her bosom, and bolted, followed by several other mothers and their offspring. The remaining "stag party," numbering some dozen, began running to and fro looking surprised and angry. I continued laughing and shouting as if I were at a football match, and soon became quite incoherent from sheer nerves. I advanced on the old male, shouting: "They've made a goal; run, run, you old idiot; bonjour, mademoiselle cochon; nunca café con leche," at the same time executing a spine-rocking rumba combined with all the other outlandish dances in my repertoire. He stopped dead in his tracks. His eyes opened wide and his whole face took on a quite ludicrously human expression. He muttered to himself. "Standing aghast" is the only way to describe his poise, as if he was just as much shocked at my behaviour as he was bewildered and frightened at what he saw. A few seconds he stood his ground, amazement written all over his face; then his nerve gave way and he shied like a dog. His final rout was accompanied by a flood of the choicest swearing from my Cockney vocabulary. He fled.

Getting There Is Half the Fun

"Look, Ma, no hands!" seems to be the daredevil brag of the three- or four-month-old baboon at left. But it must stay alert, for at any sign of danger its mother might bolt into violent action to escape, and it could fall and suffer injury or death. In its first few months a baboon infant will hug itself against its mother's breast, clinging tightly with all four limbs, and in this position it is able to ride out the most zigzagging ground race or dizzying aerial journey.

Swinging from tree to tree is not the favored mode of travel among baboons. By selective adaptation they long ago became well suited to moving on the ground. But "adaptability" is the guiding theme for the baboons' survival, and in a pinch they can execute a quite creditable leap, such as the one below.

By air and land they cover more territory foraging than any other primate. As a result, baboons are not limited to existence in forest terrain, as is the arboreal monkey, but can make their home in areas as varied as the savannas of central Africa and the sandy edges of the Sahara.

Taking a dip is not a baboon's idea of fun. It will do so reluctantly and only when there is no other way to get to where it is going. Some scientists speculate that this conditioning has come about because the baboon, a quick learner, has seen what can happen to bathing monkeys when a crocodile chances by.

As darkness begins to settle over the woodland savanna, a troop of baboons takes to the trees and settles itself in for the night. Here they will be safe except for occasional attacks by night-hunting cats. In such an event the entire troop will scamper for safety to the trees' very highest branches.

Learning Togetherness

In the world of baboons the loner is virtually unknown, for baboons are among the most social of all monkeys. The reason for their dedicated sociability: survival of the species. Baboons generally travel on the ground in open and lightly wooded areas, where they are vulnerable to attacks by predators. A lone baboon would be killed easily, but a troop can hold off almost any predator that might attack. So the impulse to "belong" has been bred into baboons by the ruthless process of natural selection and adaptation.

A baboon's training in the give-and-take rules of troop membership begins while it is still in its mother's arms. From this secure base it will venture forth on progressively more daring expeditions: a cuddle in the arms of adoring male and female adults other than its parents; a first try (left) at playing with other infants; in later months a roughhousing tumble with the dominant males, who are fierce in combat but gentle with youngsters of the troop.

An obstreperous youngster (below) looks up warily, knowing it is in for trouble, as it sees a dominant male disciplinarian approaching.

Scooping up the troublemaker, the disciplinarian inflicts punishment—a painful but nonwounding bite on the neck.

Grooming is the social act that ties together the members of a baboon troop, such as those at left. Grooming serves not only to keep a monkey's fur free of dirt and ticks but seems also to soothe even the most aggressive dominant male. Baboons of all ages get together regularly to groom and to be groomed.

Having carried out the sentence, the dominant male looks sternly at the crying youngster to see if a lesson has been learned.

Apparently satisfied, the male moves on to other matters while the contrite juvenile delinquent appears to realize the error of its ways.

Color Them Fierce

Throughout nature, color serves the survival needs of species. Seldom is this need more spectacularly served than it is for the male mandrill (left). The mandrill is a baboon and shares with other members of that simian group a frequently aggressive disposition toward its own kind. If it expressed its aggressiveness solely through combat in its climb toward troop dominance, the peace and very existence of the troop would be endangered. So the male mandrill uses its dazzling facial colors as threat signals, combining them with a penetrating stare to intimidate a male rival for power. Thus, dominance is established without injury or death.

Color plays a role in indicating submission as well. The curious object below is the nether end of a mandrill. When a male decides that an opponent facing him is tougher than he, he turns and, like most Old World Monkeys, lowers his body and meekly displays his crimson cottontail rear end to the victor.

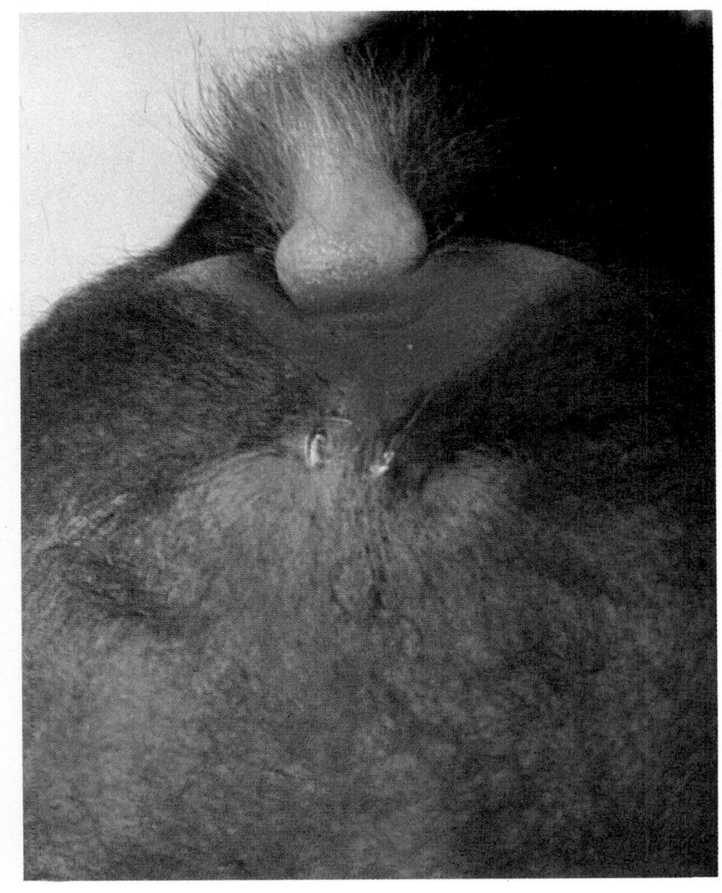

If an opponent of the Hamadryas baboon (above) were to mistake his scarlet-fanged expression for a yawn, the mistake could prove fatal, for this is a threat grimace, used by the Hamadryas to settle disputes with other males without combat. If the opponent should persist in looking for trouble, he might quickly learn that those long canine teeth are as dagger-sharp and dangerous as they appear. The threat grimace is remarkably effective. Scientists have shown a large photograph of a young male to a toothless old male, and it sent him into a corner, cringing in terror.

Macaques and Mangabeys

In numbers and geographical spread, macaques—second only to man—are the most successful primate species in the world. From Gibraltar to Japan the macaque can be found in a fantastic variety of seemingly inhospitable habitats. Smaller than the baboon—about two feet from nose to rump, weighing as much as 30 pounds—and omnivorous in habit, it scratches a living in places that other primates would shun.

The only monkey to be found in the Atlas Mountains of northwest Africa, the Barbary macaque can withstand the heat of an African summer as well as the snows of winter when the thermometer sinks to 12° F. Across the Strait of Gibraltar on the slopes of the Rock, two small colonies continue to survive with the help of rations supplied by the British Army. Though commonly called the Barbary ape, it is not an ape at all but has acquired the name because, like true apes, it has no tail.

Four thousand miles of African and Middle Eastern desert separate the Barbary apes from the rest of the macaque family. The North-West Frontier of Pakistan and Afghanistan—Bengal Lancer country—is the first outpost of the short-tailed rhesus monkey that is equally at home in the foothills of the Hindu Kush or the railway station at Benares. Hindu peoples revere the monkeys and allow them to raid crops and steal from bazaars. Rhesus monkeys take advantage of their sacred status by inhabiting temples and accepting offerings of food.

A rhesus monkey group averages 18 in number: four adult males, eight adult females and about six offspring. Among the males a hierarchy prevails. The dominant animal, usually the strongest and most aggressive, becomes the leader. He settles quarrels, organizes movements of the group and protects it from attack. In the wild, the leader always has first access to food and will not allow other group members to approach him while he is eating. He threatens anyone who gets too close by glaring, grunting or slapping the ground. If the subordinate does not promptly remove himself, he will be chased and probably bitten. To avert this, he quickly turns around and "presents" his hindquarters to the leader, who briefly mounts him.

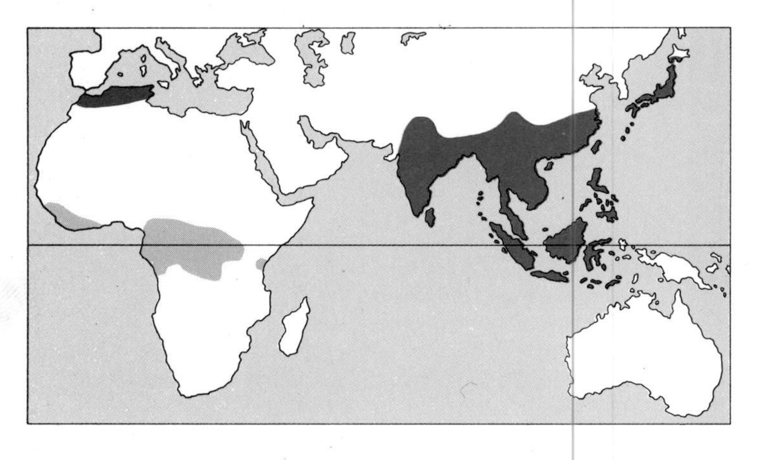

Macaques (red) live in varied habitats in Asia, the Far East, North Africa and Gibraltar. Mangabeys (blue) inhabit Africa's tropical forests.

For the female rhesus monkey the birth of an infant is an annual event. In northern India where the seasons are sharply defined by the monsoon, mating is concentrated in October, at the start of the dry season. The female shows few external signs of estrus—some reddening of the skin of the hindquarters, and occasionally her face turns an intense raspberry-red. She actively seeks to mate, presenting her hindquarters to the males until one of them joins her in a "consort" relationship, a pair-bond that lasts for a day or two, during which they mate frequently, feed, groom and sleep together. Then, after a five-and-a-half-month gestation, her infant is born in April, just before the monsoon.

While the rhesus monkey's range extends from Afghanistan to China, it is replaced farther south by the crab-eating macaque, a long-tailed beachcomber with a taste for seafood. Its favorite habitat is the mangrove swamps that fringe the coastlines of the Indonesian archipelago and the Philippines. Crabs and other small crustaceans provide a protein-rich diet. The crab-eater swims well, even underwater, but is also at home in hilly inland country.

An example of macaque adaptability is the short-tailed Japanese macaque that survives the snows of winter at the same latitude as New York. These northerners grow enormous fluffy gray coats, huddle together for warmth, strip the bark of trees for food and walk in single file through the snow. They have even learned to warm themselves by bathing in the local hot springs (opposite).

Closely related to the macaques and about the same size are the long-tailed mangabeys, which are found in Africa south of the Sahara Desert, mainly in tropical forests. Mangabeys and macaques share the ability to flatten the ears against the head, pull back the scalp and raise the eyebrows. This is usually a threat signal, and in the sooty mangabey the effect is heightened by brilliant white patches on the upper eyelids, which flash like beacons.

The Marvelous Monkeys of Koshima

From the air Koshima is undistinguished, just one more forest-covered island in the seas off the southern coast of Japan. In fact, Koshima (left) has become famous as the site of one of the most exciting and significant experiments with primates in all the world.

Underlying the purpose of this long-term experiment with the local macaque monkeys was a desire on the part of scientists to learn more about human society. They hoped that by studying the spread of new behavior patterns from an individual monkey "inventor" to other members of the group, science would gain a better understanding of how the beginnings of human culture took place.

Just how the monkeys of Koshima first got to the isolated island from their usual tropical haunts is not known, but they have been there for thousands of years. In 1953 a radical change in the monkeys' way of life was effected by the experimenters, who began to coax the monkeys to emerge from the protective forest and to gather (below) on one of the island's sandy beaches.

Today the monkeys of Koshima have changed into confirmed sun worshipers, and in the process they have demonstrated an amazing capacity to learn by observation and emulation. Most importantly, they have given scientists clues as to how human society may have gotten its start.

The Islanders' Varied Diet

"In the fall of 1953," wrote Japanese researcher Professor Masao Kawai, "a one and one half year old female, which we named Imo, one day picked up a sweet potato which was covered with sand. She dipped the potato, probably by pure accident, into water and rubbed off the sand with her hands. By this inconspicuous act, Imo thus introduced monkey culture to Koshima." Sweet potatoes had been placed at the sea's edge by the scientists in the hope of inducing the shy macaques to leave the forest for the beach, where they could be better observed. The researchers were successful beyond their wildest expectations. Soon Imo's action was copied by her playmates, her mother and her older brothers and sisters. Today, the scenes above and at left are quite routine for the Koshima monkeys as they gather at the water's edge to eat and wash their food (filmstrip at left), play, groom and sometimes even go for a swim.

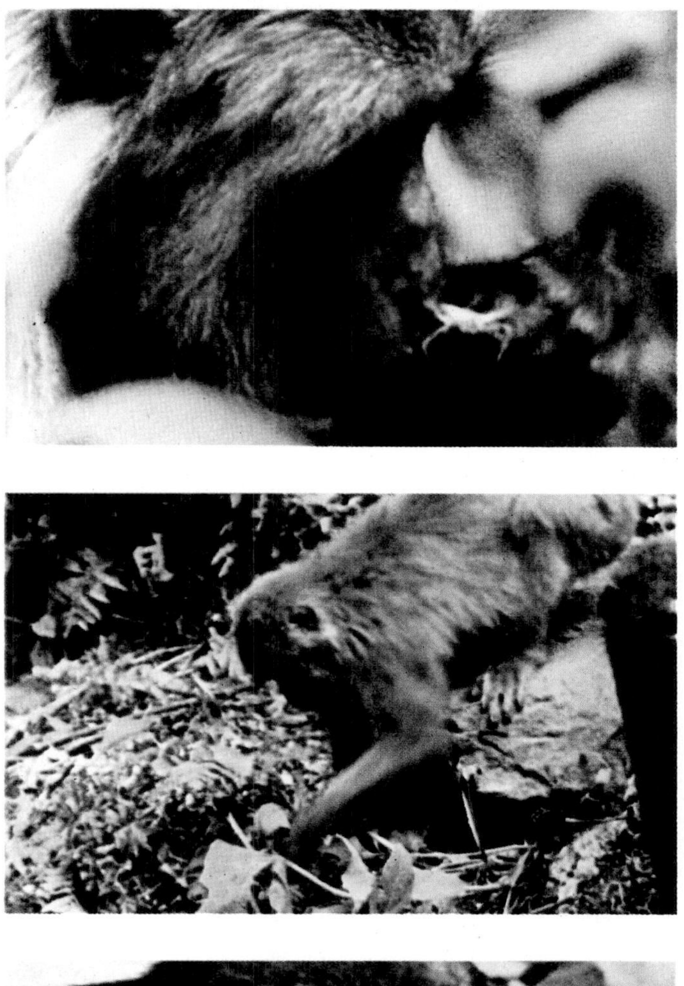

Variety in their diet is the spice of life to Koshima monkeys. They happily eat the sweet potatoes, wheat grain and peanuts supplied by the research team. But when such gourmet delights pall on the monkeys, they turn to the bountiful forest beyond the beach for a more traditional repast. There they find a tasty selection of nutritious delicacies (right, top to bottom): long-legged, crunchy crickets; insects dug out of the rich loam; nectar licked from a brilliant camellia; the leaves of tender young plants.

Aside from the unusual beach setting, unique in the world of macaques, these Koshima monkeys essentially duplicate the manners and mores of their cousins elsewhere. At top, a youngster makes a clumsy try at a headstand while a playmate and a bystander serve as dubious spectators. At center, the game suddenly changes into a wrestling match, with the victor and vanquished performing their awkward version of the mounting ritual. At bottom, a mounting ceremony allows a pair of adult males to settle a disagreement peaceably; the submissive one bends to present his rump, and the dominant male briefly mounts him in a parody of the sexual posture. This symbolic ceremony saves the aggressive macaques from constant bloody combat.

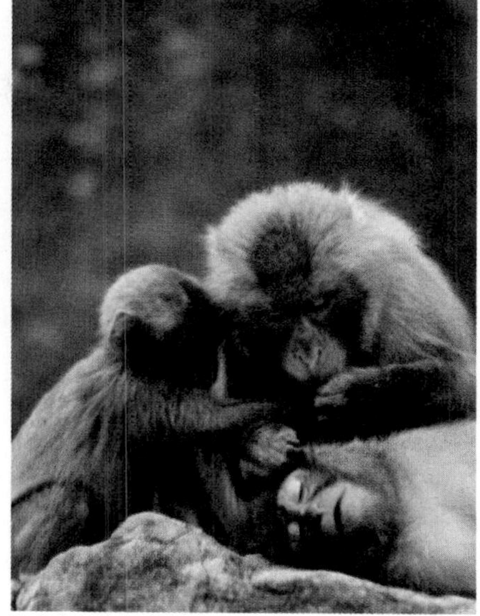

"By the sea, by the sea, by the beautiful sea . . ." It is hardly the place where you would expect to find a troop of tree-swingers, but then the Koshima macaques are hardly run-of-the-mill monkeys. They gather regularly on their private beach (left) to relax in the sun, wash their food, swim or even skin-dive to gather bits of seaweed from the ocean bottom. And while they practice their newly learned monkey-culture habits, the Koshima macaques retain many of their ancient rites. In the picture above, a mother and an older child groom a willing baby. The dejected old male below, in a scene from the Wild, Wild World of Animals *film* The Monkeys of Koshima, *has just been rejected by the females of the troop for a younger, stronger male—a fate that inevitably overtakes all dominant males who do not die before their time.*

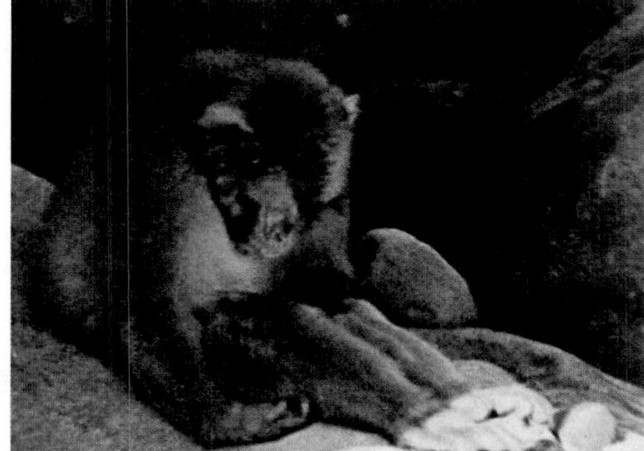

In Man's Footsteps?

The *tour de force* performances of Koshima monkeys in discovering and disseminating new behavior patterns are so intriguing to observe that there is a temptation to relate them directly to man's actions in the development of his early society. But the Japanese scientists who have been working on Koshima Island for nearly a quarter of a century caution against such facile comparisons. Macaques are monkeys after all, and their performance in each instance is necessarily a mixture of both genetic instinct and learned behavior. It will require many more years of observation of these primates before scientists can accurately distinguish between instinctual performance and true learning. So the researchers prefer not to call the monkeys' novel behavior patterns a culture but refer to it instead as pre-culture or monkey culture. Meanwhile, the Koshima troop has learned to swim and dive (filmstrip, left), to enjoy the sun and sand, oblivious to the fact that they are, in the world of primates, true superstars.

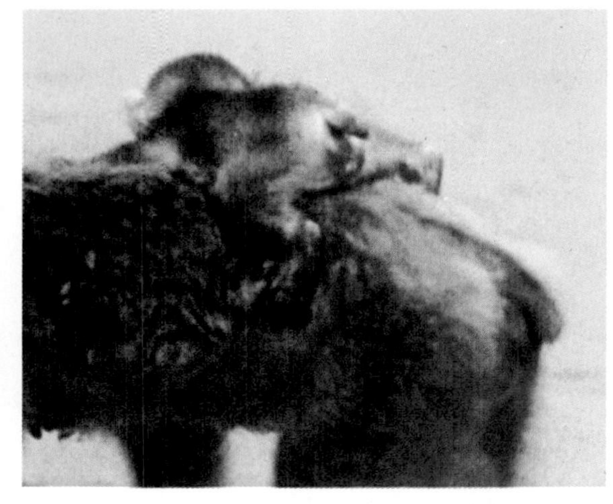

Being the subjects of serious scientific research hasn't stopped Koshima youngsters from also being happy, boisterous monkeys. Older infants (top) mimic in play the traveling position they assume on their mothers' backs (center). At an earlier age (bottom) infants hang beneath the mother, suckling or (left) submitting contentedly to grooming.

A Lion among Macaques

The shy monkey at left and below is a lion-tailed macaque, a denizen of the forests of southern India, named for its tail with a lionlike tasseled tip. There is something leonine, too, in its silky, pale-gray ruff. Living in a group, they will sometimes chase one another or leap in the air like children or bite their tails, running around in circles. When there is water nearby, lion-tailed macaques will occasionally splash in to wash themselves or to chase after any object that has caught their fancy.

Snow Monkeys of Japan

The miserable-looking little creature above, caught in a blizzard, is one of a remarkable troop of 50 snow monkeys or Japanese macaques that live in the Japanese Alps in northern Honshu (right), the northernmost habitation of any primate except man. Because of their harsh environment, this special troop has adapted to a diet of seeds, bark and other foods that no other monkey would normally touch in order to survive. Another, larger group of snow monkeys living in a slightly—but only slightly—less inhospitable habitat in the mountains west of Tokyo is shown on the following pages.

Swirling steam rises from this hot-spring pool (above), even in the midst of a driving snowstorm. Here a group of snow-monkey females and their babies relax contentedly in the salubrious waters. Naturalists report that a group of young monkeys of the troop discovered the pool and for a time had its pleasures all to themselves. But they were quickly ousted when the females of the troop discovered how wonderful a find the youngsters had made. The bitter northern climate offers few idyllic moments like this for the snow monkeys, however. Survival in their hostile environment is a difficult game, one that a number of the older or weaker members of the group lose each winter.

Snow-monkey troops are led by a dominant male who, in some cases, enjoys certain advantages—the best of the food and females—and leaves the policing of the troop to an aggressive assistant. However, the dominant male at right quite obviously believes in doing his own fighting.

Although the snow monkey's world is vastly different from that of its equatorial cousins, the habits and modes of community organization of all of them are much alike. For example, mutual grooming is as vital to the contentment and health of a snow monkey as it is to that of other monkeys. It is an act that helps bind the troop together in its continuing struggle for existence. Here, a snow-monkey mother cuddles her baby while another female grooms her.

One troop of snow monkeys lives in a frigid, snow-covered valley with the appropriate name of Jigakudani—"Hell Valley." How snow monkeys manage to survive in such a place through the six-month-long winters is under careful study by scientists from Kyoto University and the Japanese Monkey Center. They have discovered that the snow monkey can digest the frozen pulp from beneath the bark of trees—a most unsimian diet. But the search for food is a difficult and sometimes perilous one. Here a female makes a precise and acrobatic leap across a frozen river with her baby clinging tightly to her back.

Mangabey Mannerisms

The five species of mangabeys have a striking range of color, from the champagne beige of the agile mangabey munching figs (left) to the vivid contrasts of the white-collared mangabey (above). But the monkeys all have one unique feature in common—their light-colored eyelids that can be seen from great distances. It is thought that mangabeys communicate with one another by slowly or rapidly lowering and raising these remarkable lids. But, despite their fluttering eyelids, mangabeys are anything but coquettish. Although they are primarily arboreal, when they do come to the ground in cultivated areas they often wreak havoc on the local rice and cocoa crops.

The Sacred Simians

Over the centuries animals have been elevated to sacred status through various religious beliefs and national legends. Among these are two macaque monkeys, the misnamed Barbary ape (left) and the rhesus monkey (below and opposite). The Barbary ape, which was mistakenly thought to be an ape because it lacked a tail, is the only monkey found wild in Europe, where it is confined to the Rock of Gibraltar. According to popular British legend, the Rock will remain under British rule only as long as the Barbary apes are there. The tradition is so widely accepted that during World War II, when the Rock's monkey population declined, Prime Minister Winston Churchill cabled the British forces in North Africa and ordered, "Catch some monkeys for Gibraltar at once." The rhesus monkey is treated with similar reverence where the Hindus worship Hanuman, the monkey-god of magic and healing. In the West the rhesus is also associated with healing, having been used in thousands of medical experiments, notably the discovery of the blood's Rh factor.

Rhesus monkeys rarely travel alone. They prefer to wander in groups consisting of as many as 18 or 20 members. In the countries where they are worshiped, the monkeys are allowed to roam freely. They are equally at home in the trees, on the ground or even in the water (above), where they prove to be good swimmers. Rhesus monkeys have also adapted to life in cities. They can be found anywhere from the busiest commercial thoroughfares to the holiest temples, such as the one (opposite, below) in Katmandu, Nepal.

New World Monkeys

Twice a year during the rainy season the Amazon River rises punctually and dramatically, flooding hundreds of thousands of acres and thereby supporting the largest tract of tropical rain forest in the world. In the topmost branches of the forest canopy where the brilliant equatorial sun produces a luxuriant growth of leaves, flowers and fruit, South American monkeys spend their lives, largely oblivious of the gloomy swamps or jungle floor 100 or more feet below. Unlike the many terrestrial monkeys of the Old World, all of the South Americans are wholly arboreal.

Two physical characteristics distinguish the monkeys of the Old and New Worlds. New World Monkeys have flat noses with widely separated nostrils, whereas Old World Monkeys have pointed noses with nostrils set close together. But because monkey noses take many different and astonishing forms, this distinction is not always clear-cut. Neither is the second characteristic: Ischial callosities are present in many but not all Old World Monkeys but are never found in New World Monkeys. Lacking the hard, insensitive sitting pads of Old World Monkeys, New World Monkeys manage to make themselves comfortable on hard branches by crouching catlike on their hands and feet. Conversely, the prehensile tails of some New World simians do not exist in the monkeys of the Old World.

One of the largest New World Monkeys, the spider monkey, is about two feet long and has a three-foot prehensile tail. When branches are too slender for it to walk on, it simply slings itself below them and, hanging upside down by legs and tail, feeds in a relaxed and leisurely manner. When seen spread-eagled against the sky, its short bulbous body supported by five long lanky limbs, a spider monkey does resemble a gigantic tarantula.

Of all the monkeys with prehensile tails, the spider monkey's is the most highly developed. The undersurface of the last 10 inches is covered with finely ridged bare skin that provides not only a good grip but also a surface as sensitive as a fingertip. The tail is both strong enough to act as a fifth limb capable of supporting the whole weight of the animal unaided and supple enough to perform finely coordinated movements. It is astonishing to see how deftly the tip curls around an offered peanut, grasping it as delicately as a human finger and thumb would. This perfection of tail control may be some compensation for the spider monkey's lack of a thumb.

Unlike the agile spider monkey, the larger howler moves slowly and deliberately, grasping branches as a chameleon does, between its second and third fingers. Its tail seems to have a life of its own, moving independently like an octopus' tentacle, feeling for branches and lianas, wrapping itself around them, acting as an extra safety device.

The howler was one of the first monkeys to come under the scrutiny of the late Clarence Ray Carpenter, the well-known American psychologist who, in 1932, undertook the first field study of a primate and pioneered the research technique of studying behavior in the wild. What interested Carpenter particularly were the howling sessions that took place daily. The dawn chorus started about 5 A.M., reached a climax an hour later and then died down. The deep-throated guttural growling was started by a single adult male and was quickly taken up by the others, while the females added a counterpoint of short terrierlike barks. Neighboring troops responded, revealing their whereabouts and so enabling the troops to avoid one another.

In the lush vegetation of the Amazon forest, many smaller New World Monkeys are found, and in the majority tails play an important role. The capuchin, familiarly known as the "organ-grinder monkey," has a semiprehensile tail, which is in constant use as an anchor but lacks the refinement of bare sensitive skin at the tip. The little squirrel monkey's tail, though not prehensile in the true sense of the word, is a highly flexible affair. On cold nights squirrel monkeys snuggle together on a branch in a neat row, each with its tail wrapped over its shoulder for warmth. The dusky titi (opposite) entwines its tail with its mate's in a gesture of conjugal security.

The only short-tailed New World Monkey is the extraordinary uakari. The same size as a capuchin—about 18 to 20 inches—it has a six-inch bobtail that is virtually useless. With its long shaggy hair, bent back, bald head and lugubrious crimson countenance, the uakari looks like a frail old man. It isn't: Even without a prehensile tail it is one of the most sensational gymnasts in the New World.

New World Monkeys are found throughout the jungles and forests of Mexico, Central and South America.

The Excitable Spider Monkeys

The skinny, sinewy spider monkey is perfectly adapted to its treetop existence in the jungles of Mexico, Central America and South America. Its remarkable prehensile tail is capable of grasping objects and enables the monkey to dangle from a tree with all paws free to pick the fruits and nuts that are its principal food (left). When a man blunders into the territory of a troop of spider monkeys they tend to become very excited, shaking the trees, throwing fruit and broken branches and making threatening noises and gestures. But after repeated encounters with humans, the monkeys become wary, sensing that men are a threat, and retreat into the dense foliage whenever they approach.

An iron grip belies the spider monkey's fragile appearance. The spider monkey at left is firmly moored to the tree with all five "hands" in a clench that would be hard to break. Indians in Ecuador and Peru sometimes shoot spider monkeys with poisoned arrows and still lose their prey. Firmly fastened to trees with a grip that not even death will relax, the dead bodies continue to hang high in the trees, out of the reach of the hunters.

The Howler with the Remarkable Voice

Slower and more deliberate than their spider-monkey cousins, howlers rarely jump through the air but, instead, are usually seen sitting in a characteristically slumped, hunchback fashion, like the red howler at left. There are six species of howlers, which range from Mexico's Gulf Coast to southern Brazil. The celebrated howl of the male monkey was at one time thought to allure female howlers—an observation first made by Charles Darwin: The howler with the loudest mating call attracts the most females. More probably, however, the male's booming howl is a means of proclaiming his territorial rights. Aided by his oversized throat and larynx, a male howler can make himself heard for as far as three miles downwind.

Daintily sampling a leaf, a black howler monkey prepares to dine (right). Because of their special diets—certain leaves, buds and flowers from the forest, supplemented by fruits and nuts that are not easily found outside the tropics—howlers have not survived long in captivity in the past. Today, with improved diets, zoos are keeping and even breeding monkeys like the red howler above.

Two Swingers and a Straight-Arrow

The tropical forests of Central and South America seem to stretch endlessly beyond the horizon, unbroken by savannas and providing an inviting treetop world for arboreal animals. In considering this, it is easy to understand why New World Monkeys spend their entire lives high in that green world of trees. But moving from branch to branch 100 feet or more above the forest floor is a chancy business—a fall could mean injury or death—so New World Monkeys have developed specialized means of grasping and balancing. For capuchin and woolly monkeys, the answer is that remarkable prehensile tail. Titi monkeys have a straight, nonprehensile tail, but they have developed powerful, grasping feet as an additional security device.

Capuchins (above), the "organ grinder" monkeys of the past, get their name from their distinctive "caps," resembling the tonsures of Franciscan friars. Capuchins live in thick forests and move through the trees on food hunts in troops of up to 30 monkeys. Sometimes one will munch on fruit held dexterously in its tail. The woolly monkey (right) also has an agile tail, capable of supporting the animal's entire weight while it uses its hands for grooming or feeding. Woollys have huge appetites and as a result in Brazil have earned the name barrigudos, *or "big-bellies." The intertwined tails of the titi monkeys (opposite) serve as a counterbalance while they rest, hunched down, with all eight feet gripping the branch.*

The Territorial Imperative

by Robert Ardrey

Robert Ardrey caused a sensation in 1961 with his provocative book African Genesis, *in which he put forth the theory that man evolved from killer apes and owes to them his predatory instincts. In* The Territorial Imperative, *from which the following excerpts are taken, he more closely examines a facet of this theory: that territorial behavior—the staking out and protection of one's own territory—is a moving force among the world's creatures and that man has followed this evolutionary pattern in his own behavior. Here Ardrey gives an example of this drive, displayed by the callicebus or titi monkeys of South America.*

One finds in these forests no sudden, splendid tropical dawn. Here the dawn comes along like gentle, insistent fingers scratching cautiously at the nape of one's neck. Slowly the callicebus family wakes. Mother and father sleep side by side, tails frequently intertwined. He, the good husband, does all the lugging about of children, and if they have an infant under four months he will be so burdened. The family shuffles about in its heartland, its castle, its sleeping tree, lapping dew off leaves, snipping a bit of fruit or a berry or two. Then about seven o'clock, suddenly galvanized, the family makes for the periphery.

I find that one of the most touching qualities in the callicebus monkey is its willingness to sacrifice a hearty breakfast for a hearty periphery. Not unless faced by extreme emergency should I make such sacrifice myself. The little family makes no compromise with principle, but bright and early is on duty at the border, only partly fed, hankering for action, waiting for the arrival of neighbors to be angry at. Shoulder to shoulder mother and father wait, tails intertwined, nursing their grudges, feeding on their animosities, impatient for the arrival of their beloved enemies. Not one foot will the family place on the neighbors' domain unless neighbors appear, having had their dew and their scanty snack, and callicebus hell will break loose.

When I was a young man in Chicago we used to say that the secret of acknowledged Chicago vitality was the *Chicago Tribune*. We read it at breakfast, we hit the ceiling in rage either for or against it, we hit the street on a dead run, and we could not survive without it. The callicebus monkey has substituted the periphery for the *Chicago Tribune*. There is a deal of screeching to begin with. Then father intrudes. The opposing father chases him back and intrudes in turn. Now family is after family. Mothers put aside all grace and give themselves over to lifetime grudges. Juveniles learn the way of all flesh. Bedlam and bellicosity rule for half an hour or so, then someone recalls that there is another boundary undefended and unexploited. The family withdraws. The family across the way recalls that it too has another border, another enemy to become enraged at. No cards or apologies are exchanged, for the rules of the game are too well understood. Were the opponents medieval knights, haughtily bowing, spreading their mailed fists in a gesture of you-know-how-it-is, the callicebus monkey could no more perfectly execute the gallant code of chivalry. . . .

On other boundaries the contestants will oppose other

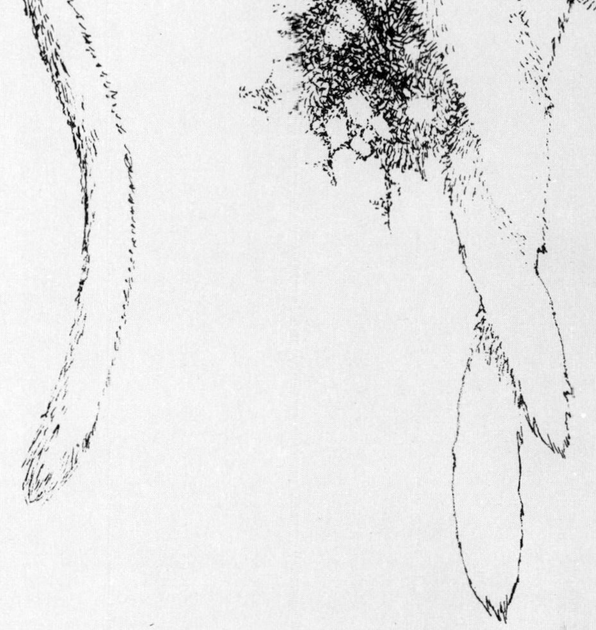

rivals. Vast must be the satisfactions of such engagements. Blood pressures rise, tissues expand, brains roil with conventional angers. Then just about nine o'clock in the morning, after a couple of hours of emotional daily dozens, it will occur to someone that somebody is hungry. That will be the end of the day's hostilities as all take their ravenous appetites to the breakfast trees. . . .

In most vertebrate species which base their social life on the pair territory, the male asserts exclusive rights over not only his space but his female. So broadly is this true of birds that, so long as the territorial concept was an ornithologist's preserve, it was generally accepted that territory was necessarily a sexual expression in the male. Until quite recently, likewise, it has been assumed that in primate species the sexual attraction of male and female has been the bond of primate society. . . . But the callicebus monkey, red-headed and white-gloved, has somehow upset both assumptions at once.

The female callicebus, unlike many primate species, has a season of heat like the lower mammals and is sexually unresponsive the remainder of the year. . . . Throughout all of that long portion of the year when sex plays no part in callicebus life, territorial defense is perfect, tolerance of intruders unthinkable, marital loyalty most estimable. Fidelity in the callicebus applies to everything, it seems, but sex. When the season of female heat arises, carnival takes over. The territorial system breaks down, borders are violated by hungering males, by famished females, and for the duration of the season the ordered animosities of the *noyau* give way to a merry-go-round of affection, a Mardi Gras of sexual adventure in the groves of unforbidden fruit. Then the season ends. Wives forgive husbands, husbands wives. All settle down to raise those inevitable bastards conceived in the *noyau*'s genetic popcorn-shaker. Side by side these marital paragons sleep, tails intertwined, on a tall branch in the dark forest. Side by side they report for duty on the periphery every morning, where, tails again intertwined, they will enjoy that more permanent of life's satisfactions, screeching at one's enemies.

Wonka, Winki and Miri

In the languages of various South American Indian tribes the words *sakiwonka, sakiwinki* and *sakimiri* all mean "small monkey," and they refer to several types of diminutive monkeys known collectively as the sakis. The uakari monkeys shown here are members of the saki family. Uakaris and true sakis closely resemble each other, with long, slender limbs, flattish faces and a timid, unaggressive manner. Their fur coloration, however, shows a seemingly limitless variation. Although they are daring jumpers, uakaris are slow, cautious climbers, testing each branch before they put their weight on it. When on the move through the hot, humid rain forests of the Amazon basin, saki monkeys travel in family units, but they will occasionally come together in greater numbers when a bumper crop of food is discovered.

The rosy-pink, hairless face of the red uakari (above) has the surrealistic appearance of a carnival mask that might be whipped off momentarily. This monkey often feeds hanging downward by its powerful legs while it carefully reaches for food with short, broad hands. Good looks do not run in the uakari family; the bald uakari (right), when bent over at rest, looks eerily like a flat-nosed old boxer dreaming of past days of glory.

Uakari and saki monkeys lack the prehensile tail that gives other New World Monkeys their special agility. But in apparent compensation, nature has given members of the saki family the most extraordinary leaping ability of any monkeys in the South American jungles. They will race upright along a high branch, arms held high for balance, then launch themselves outward (above) in a gigantic leap into the branches of a distant tree. The family unit consists of a male, a female and an infant (left). The male, usually unassertive, can turn quite aggressive if his mate or baby is threatened.

The saki ranks among the least monkeylike of all monkeys. Its face resembles that of an old man's, its voice sometimes sounds like a bird's cry, its teeth are similar to a cat's and its bushy tail to a fox's. When a saki jumps from tree to tree (above) its tail flies out behind, a dramatic pennant that probably serves to steady its flight. Naturalists have little knowledge of the habits and social customs of saki monkeys in the wild, because they are easily panicked and field studies of them are difficult. But it is known that they have omnivorous appetites. The saki at left is stripping berries from a small twig into its mouth. Sakis also eat nuts, leaves, insects, small rodents, wild grains and even birds and bats.

The High-flying Squirrel Monkeys

Tail held high and head pointed toward the ground far below, a squirrel monkey (above) dangles from a thread-thin branch, demonstrating its virtuosity as a gymnast. The small, delicate squirrel monkeys live in the high canopy of tropical forests through most of Central and South America. They follow "paths" through the trees when on the move and descend to feed at the forest's edge and along rivers, where berries, nuts and fruits grow in abundance. But they will not venture down to the ground unless they are in a very large troop. Squirrel monkeys are immaculately clean animals, despite the fact that they rub their fur, and particularly their tails, with a musky-smelling glandular secretion, which turns away hunters who might otherwise kill them for food. Some scientists believe that the purpose of this action is to leave an odor-marked trail for other monkeys to follow as the troop goes through the trees.

Night Owl

When the sun goes down and the other New World Monkeys bed down for the night, the bright-eyed, bushy-tailed douroucouli gets up and begins to prowl the rain forest in search of fruit, insects, birds' eggs and bats. The only nocturnal monkey in the world, it is also known as the night monkey or, because of its large eyes, as the owl monkey. Douroucoulis travel and hunt in pairs or in small family groups and live, like some owls, in hollow trees. They are distinguished by chalk-white faces, with dramatic black widow's peaks.

Encounters with Animals

by Gerald Durrell

Gerald Durrell, born in India in 1925, spent much of his childhood traveling with his family. Even as a young boy Durrell had an interest in local animals wherever he lived and often kept a variety of them as pets. After serving as a student zookeeper for the Whipsnade Park Zoo in Bedfordshire, England, he made various animal-collecting expeditions. During an expedition in Guyana he was able to observe two species of New World Monkeys. The following portrait of the douroucouli and the sketch appearing on page 125 are from Encounters with Animals, *one of more than a dozen books Durrell has written. The portrait reflects the affection as well as the objectivity that Durrell was able to bring to his observations.*

Another fascinating creature that used to come to the fruit trees was the douroucouli. These curious little monkeys, with long tails, delicate, almost squirrel-like bodies and enormous owl-like eyes, are the only nocturnal species of monkey in the world. They arrived in small troops of seven or eight and, though they made no noise as they jumped into the fruit trees, you could soon tell they were there by the long and complicated conversation they held while they fed. They had the biggest range of noises I have ever heard from a monkey, or for that matter from any animal of similar size. First they could produce a loud purring bark, a very powerful vibrating cry which they used as a warning; when they delivered it their throats would swell up to the size of a small apple with the effort. Then, to converse with one another, they would use shrill squeaks, grunts, a mewing noise not unlike a cat's and a series of liquid, bubbling sounds quite different from anything else I have ever heard. Sometimes one of them in an excess of affection would drape his arm over a companion's shoulder and they would sit side by side, arms round each other, bubbling away, peering earnestly into each other's faces. They were the only monkeys I know that would on the slightest provocation give one another the most passionate human kisses, mouth to mouth, arms round each other, tails entwined.

Marmosets

Although they are found only in the tropical forests of the Western Hemisphere, marmosets occupy their own branch of the anthropoid family tree and are classified separately from all the other New World Monkeys, the cebids. What sets the marmosets apart is the fact that they alone among the anthropoids are equipped with claws rather than nails—except for a flat nail on their big toes—and they are the smallest of the monkey family. The pygmy marmoset is not much larger than a mouse and can be held in the palm of the hand.

But most marmosets are about the same size as small squirrels and show the same squirrel-like agility and shyness as they dodge behind a tree trunk to escape observation and then peep furtively around to satisfy their curiosity. Their long curved claws give them an excellent grip on rough bark as they scamper up vertical trunks or run along big branches.

Marmosets are found in almost all forested areas of Central and South America from Panama to southern Brazil. The two main groups, the true marmosets and the tamarins, are basically very similar but have a remarkable assortment of markings and adornment. Most marmosets sport conspicuous ear tufts—white tassels or black plumes or short yellow sprouts—but some have completely bare ears of a delicate pink shade that flush scarlet in moments of agitation. Some tamarins have mustaches that range from a neat white blob on the muzzle to the long "handlebars" of the emperor tamarin, named after the Emperor Franz Josef of Austria, whose magnificent mustaches were legendary. In addition, some have crests on the head; the pinché, or "cottontop," tamarin, for instance, has fantastic white plumes like the warbonnet of an Indian chieftain.

Like the gibbons, marmosets live in a "family" social group consisting of a mated pair and their offspring. After a gestation of about five months, usually twins are born which the mother soon hands to the father to look after. Most of the time he carries them about on his back, but at feeding times he huddles close to the female, allowing the babies to scramble onto her to nurse. At this stage they are tiny and very little trouble; but they grow fast, and by the time they are two months old the father looks weighed down by his responsibilities. The adolescent members of the family carry the babies part of the time, and thus the rudiments of baby care are learned at an early age. Sharing the burden of parenthood is obviously of value, particularly with twins. If the mother had to carry as well as feed them, it would be an intolerable strain. By the time the young reach the age of six months they are fairly independent, and their mother may once again have given birth.

Male and female marmosets are of equal size and status within the group. The father is dominant over his sons, the mother over her daughters. As with gibbons, the maturing young are subjected to increasing antagonism from one of their parents and eventually quit the group to start families of their own.

Marmosets have fewer facial expressions than other monkeys, but they can convey information as well as emotion in other effective ways. A long plaintive whistle helps to keep the family in touch with one another even when visual contact is lost. High-pitched trills are used to give the alarm when a predatory bird flies overhead.

In threat display a marmoset flattens its ear tufts, lowers its eyebrows and fluffs up its fur, making itself look incredibly ferocious—that is, to another marmoset. Scent signals are very important, too, though their significance is lost on a human observer. A marmoset displays dominance over another marmoset in a ritualistic gesture by backing toward a subordinate with tail erect and hair fluffed out, forcing it to become aware of the overpowering scent emanating from the glands of its genital region.

It is a surprising truth that man—one of the largest anthropoids—coincidentally shares with the smallest of them three very significant characteristics: the same type of family social group, with the father playing a major role in the upbringing of the offspring; the same number of teeth—32; and the same number of chromosomes—46.

Of course, this is just coincidence. And yet, looking into a marmoset's wide inquisitive eyes and its intelligent little face, a miniature in so many ways of those of human beings, it is impossible not to feel a sense of kinship with the smallest of our monkey relatives.

Marmosets live in the upper strata of the Central and South American rain forests.

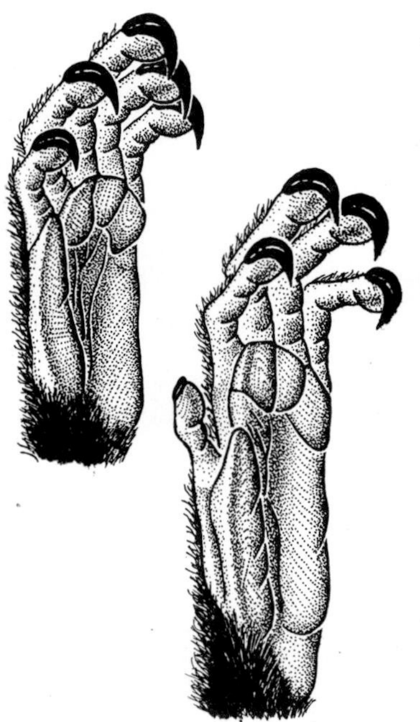

Unlike other monkeys, marmosets and tamarins have long, curving nails on all the digits of their hands (above, left) and feet (above, right) except their big toe. Their thumbs are not opposable, but for these lightweight simians claws provide an adequate means of holding on as they run along the branches of a tree.

The Smallest of All

The diminutive marmosets and tamarins dwell in the uppermost strata of the rain forests of Central and South America. These omnivorous monkeys rarely leave their treetop homes, since the fruit, nuts, buds, insects, birds' eggs and small lizards they live on are available there in abundant supply. The short-limbed marmosets are excellent climbers. Their sharp claws allow them to get a firm grip on branches and tree trunks so that even when they are on the lookout for potential danger, like the two cottontop marmosets above, or are being poked and prodded by an inquisitive relative, such as the white-fronted marmosets at left, there is little chance of their falling. The slightly larger, longer-limbed tamarins, like the sublimely mustachioed emperor tamarin (opposite), are not only good climbers but accomplished jumpers as well—a talent that gives them extra protection against such predators as the eagle and man.

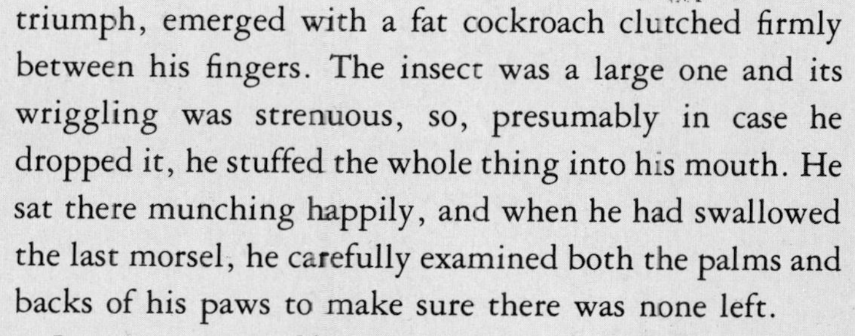

Encounters with Animals *by Gerald Durrell*

One of the smallest of these animals was Pavlo, a black-eared marmoset, and his story really started one evening when, on a collecting trip in British Guiana, I sat quietly in the bushes near a clearing, watching a hole in a bank which I had good reason to believe contained an animal of some description. The sun was setting and the sky was a glorious salmon pink, and outlined against it were the massive trees of the forest, their branches so entwined with creepers that each tree looked as though it had been caught in a giant spider's web. There is nothing quite so soothing as a tropical forest at this time of day. I sat there absorbing sights and colours, my mind in the blank and receptive state that the Buddhists tell us is the first step towards Nirvana. Suddenly my trance was shattered by a shrill and prolonged squeak of such intensity that it felt as though someone had driven a needle into my ear. Peering above me cautiously, I tried to see where the sound had come from: it seemed the wrong sort of note for a tree-frog or an insect, and far too sharp and tuneless to be a bird. There, on a great branch about thirty feet above me, I saw the source of the noise: a diminutive marmoset was trotting along a wide branch as if it were an arterial road, picking his way in and out of the forest of orchids and other parasitic plants that grew in clumps from the bark. As I watched, he stopped, sat up on his hind legs and uttered another of his piercing cries; this time he was answered from some distance away, and within a moment or two other marmosets had joined him. Trilling and squeaking to each other, they moved among the orchids, searching diligently, occasionally uttering shrill squeaks of joy as they unearthed a cockroach or a beetle among the leaves. One of them pursued something through an orchid plant for a long time, parting the leaves and peering between them with an intense expression on his tiny face. Every time he made a grab the leaves got in the way and the insect managed to escape round the other side of the plant. Eventually, more by good luck than skill, he dived his small hands in amongst the leaves and, with a twitter of triumph, emerged with a fat cockroach clutched firmly between his fingers. The insect was a large one and its wriggling was strenuous, so, presumably in case he dropped it, he stuffed the whole thing into his mouth. He sat there munching happily, and when he had swallowed the last morsel, he carefully examined both the palms and backs of his paws to make sure there was none left.

I was so entranced by this glimpse into the private life of the marmoset that it was not until the little party had moved off into the now-gloomy forest that I realized I had an acute crick in my neck and that one of my legs had gone to sleep.

It is a curious thing, but when you keep animals as pets you tend to look upon them so much as miniature human beings that you generally manage to impress some of your own characteristics on to them. This anthropomorphic attitude is awfully difficult to avoid. If you possess a golden hamster and are always watching the way he sits up and eats a nut, his little pink paws trembling with excitement, his pouches bulging as he saves in his cheeks what cannot be eaten immediately, you might one day come to the conclusion that he looks exactly like your own Uncle Amos sitting, full of port and nuts, in his favourite club. From that moment the damage is done. The hamster continues to behave like a hamster, but you regard him only as a miniature Uncle Amos, clad in a ginger fur-coat, for ever sitting in his club, his cheeks bulging with food. There are very few animals who have characters strong and distinct enough to overcome this treatment, who display such powerful personalities that you are forced to treat them as individuals and not as miniature human beings. Of the many hundreds of animals I have collected for zoos in this country, and of the many I have kept as pets, I can remember at the most about a dozen creatures who had this strength of personality that not only made them completely different from others of their kind, but enabled them to resist all attempts on my part to turn them into something they were not.

olden lion marmosets

Credits

All photographs in this book are by Nina Leen except those listed below:

Cover photograph—Co Rentmeester, T-L P.A. 1—George Holton from Photo Researchers, Inc. 5—Larry Burrows, T-L P.A. 6—(bottom, left) Co Rentmeester, T-L P.A. 6–7—(top) Ralph Morse, T-L P.A., (bottom) Stephen Zoloth. 7—(top, right) Stanley Washburn, T-L P.A., (bottom, right) Helmut Albrecht from Bruce Coleman, Inc. 15—Bruce Coleman, Inc. 16—Ph. Grossa from Jacana. 17—Tom McHugh from Photo Researchers, Inc. 18—(top, left) Ph. Grossa from Jacana, (bottom, right) Lee Lyon from Bruce Coleman Ltd. 18–19 (top) Lee Lyon from Bruce Coleman Ltd. 19—(right) Lee Lyon from Bruce Coleman, Inc. 20–21—Dian Fossey from Bruce Coleman, Inc. 21—Lee Lyon from Bruce Coleman, Inc. 27—Wolfgang Bayer. 28—(left) Jacques Jangoux from Photo Researchers, Inc. 28–29—Karen Minkowski. 30—Helmut Albrecht from Bruce Coleman Ltd. 31—(top) Helmut Albrecht from Bruce Coleman Ltd., (bottom) Wolfgang Bayer. 36—Helmut Albrecht from Bruce Coleman, Inc. 37—Helmut Albrecht from Bruce Coleman Ltd. 38–39—Helmut Albrecht from Bruce Coleman Ltd. 44—(top, left) Co Rentmeester, T-L P.A., (bottom, left) Co Rentmeester, T-L P.A., (bottom, right) Wolfgang Bayer. 45—(top, left) Michael Rougier, T-L P.A., (bottom, left) Co Rentmeester, T-L P.A. 46—(top) Co Rentmeester, T-L P.A., (bottom) Wolfgang Bayer. 47—(top, left) Wolfgang Bayer, (top, right) Wolfgang Bayer, (bottom, left) Co Rentmeester, T-L P.A. 50—(top) Michael Rougier, T-L P.A., (bottom) Co Rentmeester, T-L P.A. 50–51—Co Rentmeester, T-L P.A. 51—Co Rentmeester, T-L P.A. 58—George Silk, T-L P.A. 59—(bottom) E. H. Rao from Natural History Photographic Agency. 62—(bottom) E. H. Rao from Natural History Photographic Agency. 63—(top and bottom) Alfred Eisenstaedt, T-L P.A. 64—Carlo Bavagnoli, T-L P.A. 65—Bill Ray, T-L P.A. 66—Larry Burrows, T-L P.A. 67—Co Rentmeester, T-L P.A. 70—(right) Bruce Coleman, Inc. 72—R. L. Fleming from Bruce Coleman, Inc. 73—(top) Brian Hawkes from Natural History Photographic Agency, (bottom) Andrew Anderson from Natural History Photographic Agency. 75—Bill Ray, T-L P.A. 76—Bill Ray, T-L P.A. 77—(top) George Holton from Photo Researchers, Inc., (bottom) Bill Ray, T-L P.A. 80—(top) Nadine Zuber from Rapho/Photo Researchers, Inc., (bottom) R. D. Estes from Photo Researchers, Inc. 81—(top, left) Christina Loke from Photo Researchers, Inc., (right) Alfred Twoomey from Photo Researchers, Inc. 82—(top) Stanley Washburn, T-L P.A., (bottom, left and right) Irven DeVore, T-L P.A. 83—(top) Brian Hawkes from Natural History Photographic Agency, (bottom, left and right) Irven DeVore, T-L P.A. 84—Toni Angermayer from Photo Researchers, Inc. 85—(top, right) Toni Angermayer from Photo Researchers, Inc. 87—Co Rentmeester, T-L P.A. 88–89—Steven Green. 89—Steven Green. 90–91—Stephen Zoloth. 92–93—Stephen Zoloth. 93—(top, right) Karen Minkowski. 94–95—Alfred Eisenstaedt, T-L P.A. 96—(top) Karen Minkowski, (bottom) E. H. Rao from Natural History Photographic Agency. 97—(left) Co Rentmeester, T-L P.A., (right) Wolfgang Bayer. 98–99—Co Rentmeester, T-L P.A. 99—Co Rentmeester, T-L P.A. 100—Co Rentmeester, T-L P.A. 101—Co Rentmeester, T-L P.A. 102–103—Lysa Leland. 103—Bruce Coleman, Inc. 104—(top) R. J. Johns from Bruce Coleman, Inc., (bottom) M. N. Boulton from Bruce Coleman, Inc. 105—Ivan Polunin from Natural History Photographic Agency. 107—Bruce Coleman, Inc. 108—(bottom, left) E. H. Rao from Natural History Photographic Agency. 110—(bottom, right) Bruce Coleman, Inc. 114—(left) Milwaukee from Jacana.

Photographs on endpapers are used courtesy of Time-Life Picture Agency and Russ Kinne and Stephen Dalton of Photo Researchers, Inc.

FILM SEQUENCES and individual frames are from "The Monkeys of Koshima," "Green Ceilings of Borneo," "Man's Closest Kin" and "Links to Man's Past," programs in the Time-Life Television series *Wild, Wild World of Animals*.

MAPS on pages 14, 26, 42, 52, 60, 68, 74, 86, 106 and 120 are by Enid Kotschnig.

ILLUSTRATIONS on pages 9, 12 and 122 are by Lorraine J. Meeker and Chester S. Tarka. Evolution chart on pages 10–11 is by Peter Barrett.

Other illustrations on pages 22–23, 25 and 29 are by André Durenceau. Illustrations on pages 32–35, 40–41, 48–49, 78–79, 112–113, 119 and 125 are by Jennifer Perrott. Illustration on page 57 is by Tony Chen.

Bibliography

NOTE: Asterisk at the left means that a paperback volume is also listed in *Books in Print*.

Amon, Aline, *Reading, Writing, Chattering Chimps*. Atheneum, 1975.

Bourne, Geoffrey H., *Primate Odyssey*. G. P. Putnam's Sons, 1974.

Chance, Michael R., and Jolly, Clifford, *Social Groups of Monkeys, Apes and Men*. E. P. Dutton, 1971.

DeVore, Irven, ed., *Primate Behavior: Field Studies of Monkeys and Apes*. Holt, Rinehart and Winston, 1965.

Eimerl, Sarel, and DeVore, Irven, *The Primates*. Time-Life Books, 1965.

Grzimek, Bernhard, *Grzimek's Animal Life Encyclopedia, Volume 10, Mammals I*. Van Nostrand Reinhold, 1972.

Harrisson, Barbara, *Orang-utan*. Doubleday, 1963.

Hill, W. C., *Primates: Comparative Anatomy and Taxonomy*. John Wiley & Sons, 1955.

Jolly, Alison, *The Evolution of Primate Behavior*. Macmillan, 1972.

Mackinnon, John, *In Search of the Red Ape*. Holt, Rinehart and Winston, 1974.

Morris, Desmond, ed., *Primate Ethology*. Anchor Books (paperback), 1967.

*———, *The Naked Ape: A Zoologist's Study of the Human Animal*. McGraw-Hill, 1968.

Napier, John, and Napier, Prue, *Old World Monkeys: Evolution, Systematics and Behaviour*. Academic Press, 1970.

———, *A Handbook of Living Primates*. Academic Press, 1967.

Napier, Prue, *Monkeys and Apes*. Bantam Books (paperback) 1973.

Reynolds, Vernon, *The Apes: Gorilla, Chimpanzee, Orangutan and Gibbon*. E. P. Dutton, 1967.

Rowell, Thelma, *The Social Behavior of Monkeys*. Penguin Books (paperback) 1973.

Schaller, George B., *Mountain Gorilla: Ecology and Behavior*. University of Chicago Press, 1963.

*———, *The Year of the Gorilla*. University of Chicago Press, 1964.

Schultz, Adolph H., *The Life of Primates*. Universe Books, 1969.

*Van Lawick-Goodall, Jane, *In the Shadow of Man*. Houghton Mifflin, 1971.

Walker, Ernest P., *Mammals of the World*. 3 vols. Johns Hopkins University Press, 1975.

Williams, Leonard, *Samba and the Monkey Mind*. W. W. Norton & Co., 1965.

Index